SLOW DISTURBANCE

DISTUR

Infrastructural Mediation

Sign, Storage, Transmission
A SERIES EDITED BY JONATHAN STERNE AND LISA GITELMAN

SLOW
BANCE

on the Settler Colonial Resource Frontier   Rafico Ruiz

Duke University Press

DURHAM AND LONDON   2021

Designed by Courtney Leigh Richardson
Typeset in Chaparral Pro and Montserrat by Copperline Book Services

Library of Congress Cataloging-in-Publication Data
Names: Ruiz, Rafico, [date] author.
Title: Slow disturbance : infrastructural mediation on the settler
colonial resource frontier / Rafico Ruiz.
Other titles: Sign, storage, transmission.
Description: Durham : Duke University Press, 2021. | Series: Sign, storage,
transmission | Includes bibliographical references and index. Identifiers:
LCCN 2020027615 (print)
LCCN 2020027616 (ebook)
ISBN 9781478007982 (hardcover)
ISBN 9781478008507 (paperback)
ISBN 9781478012139 (ebook)
Subjects: LCSH: Fisheries—Newfoundland and Labrador—History—19th
century. | Fisheries—Newfoundland and Labrador—History—20th
cen-tury. | Colonists—Newfoundland and Labrador—History—19th
century. | Colonists—Newfoundland and Labrador—History—20th
century. | Newfoundland and Labrador—Economic conditions.
Classification: LCC SH229.R859 2021 (print) | LCC SH229 (ebook) |
DDC 333.95/609718—dc23
LC record available at https://lccn.loc.gov/2020027615
LC ebook record available at https://lccn.loc.gov/2020027616

Cover art: Stills from Grenfell Mission found footage from the 1930s to the
1950s. Courtesy the Labrador Institute Archives, Memorial University.

This book received a publication subsidy from Duke University Press's First
Book Fund, a fund established by Press authors who donated their book
royalties to help support innovative work by junior scholars.

To Paulina, for teaching me not to read

ACKNOWLEDGMENTS

The Grenfell Mission story is one that can be told and retold and, with each new telling, informs the ways in which this island and deep coast are and have been connected to the world. The lands of the Beothuk, Mi'kmaq, Innu, and Inuit demand recognition as past, present, and future homes. They precede, underlie, and should take precedence over these connections to other lands and stories. It is for that reminder of connection that I thank the numerous residents I spoke to across northern Newfoundland and Labrador. And it was, largely, as "residents," people with a long experience in situ, both Indigenous and settler, that they helped me to understand the slow and ongoing evolution of the International Grenfell Association. These conversations made the many hours spent in archives feel like a life in their own pasts and that I was privileged to be able to make that trip. My sincere thanks in particular to Dorothy McNeil for sharing her father's remarkable legacy, one that could lie just below the surface, and also to Agnes and Francis Patey for making the past inhere in small but dear objects. Thanks as well to Jane McGillivray and the Michelin family in North West River, Labrador, for inviting me in with so much kindness and generosity.

A book about infrastructure of course needs to start by thanking those who constituted its own infrastructural community of a kind. To my wife and partner, Paulina Mickiewicz, for being the scholar I admire most and for creating her own path toward everyday fulfillment. Loving thanks as well to Anabel and Lola, who came into being while I wrote this book and, in that being, have made the world both smaller and larger, reminding me of the futures that are just around the corner for them. Finally, to my family, thanks for sticking with the eternal student. Immeasurable thanks to my mother, Janice McAuley, for her help in looking to those futures. My second parents, Bogda and Rafał Mickiewicz, thank you for never losing trust in the accomplishment. Sisters and brothers, by birth and law, Aly Ruiz, Simon Ruiz, Michael Tsourounis, Heather Maxted, Magdalena Mickiewicz, Rafał Mickiewicz (jr),

Lukasz Lewandoski, and Luisa DiStefano, all of you made coming home wherever you all were. Y gracias a Martha Zalazar—yo siempre podia llamarte.

In making the leap from dissertation to the book you hold in your hands (or that waves in front of your eyes in bands of light), there are so many helping hands and minds to acknowledge and thank. It took shape in between McGill University's School of Architecture and the Department of Art History and Communication Studies, in that odd, self-inflicted purgatory known as an "ad personam" PhD. Beyond the administrative hurdles, what made it worthwhile and held it together were the friends I made along the way. So many departmental, scholarly, and not-so-scholarly friends made the "dissertation years" a true time of learning—that sense of "making," of being around, here and there, over so many years—and deserve a "thank you, friends." You are so many and so wonderful.

I also feel incredibly fortunate to have found a friend and mentor by simply being curious and wondering myself into a PhD. Darin Barney, my doctoral cosupervisor, went above and beyond his supervisory duties to ensure that the work got done and that he read every word of it. He is the most relentlessly generous scholar I know and someone who helps "infrastructural worlds" to grow through community building and trust. Thanks as well to my cosupervisor Annmarie Adams. She gave me an intellectual home in the school without thinking twice, for which I will always be grateful, and in the process taught me that the material matters and that it deserves the captions we give it. My committee was also an early momentum builder that helped me see that there was indeed a book on the horizon. For taking on a long dissertation in a time of short ones, my heartfelt thanks to Alessandra Ponte, Jonathan Sterne, Will Straw, and Ipek Türeli. Thanks as well to my external reviewer, Jody Berland, who was the first substantive reviewer of this project, looking in from outside the friendly confines of one's department and university, and seeding the merit that was buried in the project through the sharing of generous and substantive thoughts. It was a real encouragement at the time and one that still resonates in these pages. I also feel privileged to have learned both with and from caring and critical scholars such as Carrie Rentschler, Jenny Burman, and Andrew Piper. And for getting me from beginning to end at McGill, with only a few administrative misdirections of my own doing, my gratitude goes to Maureen Coote and Wambui Kinyanjui.

During the writing of this book I was fortunate to meet actually quite rare creatures in our scholarly world—honest, generous, and critical readers.

So my infinite thanks to Melody Jue, Shannon Mattern, Amy Slaton, Florian Sprenger, Nicole Starosielski, and Hannah Tollefson for delving into the Grenfell-verse when it needed direction and a helping hand.

For helping me find things and learn about them, my thanks to Janet Parks at the Avery Architectural and Fine Arts Library; Renata Guttman and Colin MacWhirter at the library of the Canadian Centre for Architecture; Jack Eckert, Jessica Murphy, and the other archivists at the Countway Library of Medicine; Morgon Mills at the Labrador Institute Archives; Nora Hague, Céline Widmer, and the whole team at the McCord Museum and Archives; the archivists at the British Postal Museum and Archives, the National Archives at Kew, the Rooms Provincial Archives of Newfoundland and Labrador, and the Sterling Library at Yale University. For helping me get places and facilitating that still so important sense of "being there," even or especially if that "there" is a distant archive, my thanks to the Countway Library of Medicine, the Faculties of Arts and Engineering at McGill University, the Institute for the Public Life of Arts and Ideas at McGill University, the McCord Museum, Media@McGill, McGill University's School of Architecture, the Smallwood Foundation of Memorial University, the Social Sciences and Humanities Council of Canada, and Gabriella Coleman, the Wolfe Chair in Scientific and Technological Literacy. Thanks as well to Paul Canning and the International Grenfell Association for permission to reproduce archival materials.

Various versions of the chapters that follow have been presented at conferences, worked over at workshops, and listened to at talks. My thanks to those so often overlooked yet peopled categories of the "organizer" and the "engaged participant"—you have made traveling to so many places so worthwhile. In particular, I'd like to thank Joanna Zylinska and the Department of Media and Communications at Goldsmiths for the opportunity to spend some much-needed thinking time with them in London. Joanna helped "mediation" mean (along with Sarah Kember, of course). Thanks as well to Armin Beverungen and the whole intellectual community at the University of Lüneburg's Digital Cultures Research Lab. They welcomed Paulina, Anabel, and me in with such open and smart arms that we didn't want to leave.

Finally, my thanks to Courtney Berger at Duke for persevering with this project over the years. It has been a true collaboration that Courtney has cultivated with vision and purpose. Thanks as well to Sandra Korn and the whole DUP team for giving such value to ink, paper, and ones and zeros—I feel so privileged to have worked with all of you. Jonathan Sterne and Lisa

Gitelman, thanks for bringing me into the critical space of the Sign, Storage, Transmission series—this book could be nowhere else. As per convention, two anonymous reviewers were the not-so-silent voices that shaped this book in substantive and profound ways. I am immensely grateful for their synthetic comments and willingness to see the infra- in the structure.

INTRODUCTION. FIRST FISH,
THEN MEDIATION

FIGURE I.1. Detail of the ceramic mural designed by Jordi Bonet for the rotunda of the Charles S. Curtis Memorial Hospital, which opened in St. Anthony in 1966. St. Anthony, Newfoundland and Labrador, 2011. Photograph by author.

The fish came first. It is a bright November day on the northernmost tip of the island of Newfoundland. Wave upon wave ferries in on the rocky coasts that contour where land ends and the North Atlantic rises. At our back is L'Anse aux Meadows. Jellyfish Cove. That marker of the fallibility of the Atlantic. Of Norse ambition. Of the promise of colonial contact to come. Farther back still, closer to the geometric and geological center of the triangularly shaped island, is Gros Morne, a plateau that was once the bottom of the Iapetus Ocean. The Proto-Atlantic. It is a place of very old bivalves, corals, cephalopods. A place of inscription and first fish. From the peak of the plateau you look out over gray-green elemental water. Ocean to ocean. Fish of the past and fish of the future. Paleontology. Fishery. This is the promise of extraction.

FIGURE I.2. Spoon, saw, and scissors found at a former Beothuk site on the Exploits River; the Beothuk would rework tools retrieved from settler outport fishing communities to serve as part of their hunting tools, including arrowheads and harpoon tips. "The Beothuk," Newfoundland and Labrador Heritage, www.heritage.nf.ca /articles/aboriginal/beothuk.php.

The fish came first. The Beothuks were refugees in the making. With the arrival of European colonial settlement in the seventeenth century, the Beothuks were forced to cede their coastal modes of living to British and French settlers. They had to go from bone harpoons to arrow points. A forced departure from sea to land. Fish to the Beothuk were a precontact world of life. The opposite of extinction.

The fish came first. For North Atlantic fisherfolk involved in undertaking late nineteenth-century fishery, this was an existential decision. Off the coast of the island of Newfoundland and up into the Labrador Sea, thousands of settler colonists would stake their claims in the resource boom-and-bust economy that was part of the lifeblood of Britain's oldest colony. This New Found Land, the easternmost edge of the North American continent, was part of the first wave of European colonization, and by the turn of the twentieth century, it was a full-fledged resource frontier operating across the spectrum of colonial capitalism. The settler fisherfolk, having supplanted the Beothuk by

FIGURE I.3. "Big cod fish from the trap, Battle Harbour, Labrador." Photograph from Holloway, *Through Newfoundland with the Camera*.

the early nineteenth century through forced expulsions, the transmission of disease, and other means of colonial settlement, were engaged in a mercantile system that privileged the extraction, processing, and distribution of fish, largely *Gadus morhua*; their specie was Atlantic cod. Sun, salt, and air had to give way to bone-dry fish. This was their promise of extraction.

———————

THE FISH CAME FIRST. Fossil, food, commodity. Natural history, ontology, economy. The promise of extraction. This is a book about the making and breaking of that promise. I begin with the afterlives of North Atlantic fish as a means of reflecting on how resource frontiers get made and unmade through what I call *infrastructural mediation*. *Mediation* is certainly a complex and emergent conceptual terrain. Scholars across media studies and beyond have sought to assess how it both divides and connects "channels and protocols";[1] merges and remediates old and new media forms;[2] is at the heart of the "genesis of the media concept," particularly across the philological record;[3] and encompasses a vertical field of control at a distance.[4] Mediation is an inferred process that appears across practices of communication. It high-lights how such practices come into being and what ontological, epistemolog-ical, and material ground is both covered and bound together through such practices. I anchor *mediation* in environmental media studies and attempt to bring it to bear on current debates surrounding colonial forms of environ-ment making that emerge across the sites of extractive capitalism. "Frontiers aren't just discovered at the edge," as Anna Tsing has it; "they are projects in making geographical and temporal experience."[5] Infrastructural mediation designates a process that attends to the materialization of infrastructural ar-rangements across past, present, and future colonial lifeworlds, with particu-lar attention to the contested sites created by extractive capitalism.

Infrastructures, as Lauren Berlant suggests, are not rigid systems that structurally condition social formations. Rather, they are defined by their "patterning of social form," their capacity to account for the emergence of certain organizations of life over others.[6] Across the Global North and South, for all the talk of our depleted planet, the project of resource frontier mak-ing recursively returns, from ocean floor to glacial peak. Resources are seem-ingly always and everywhere made to appear, though predominantly across the precarious ecosystems that have been at the margins of historical zones of extraction, through practices of communication understood as an infra-structural condition of transportation and a more general political economy of capitalist logistics. The processes of extraction that characterize particu-lar appropriated environments are akin to the mediating properties of con-ventional media technologies. The protocol of sender-signal-receiver is anal-ogous to and, given our planetary condition of carbon saturation, superseded by processes of extraction, commoditization, and distribution—a resource frontier is an extractive medium. Infrastructural mediation tracks the mobile and world-creating material politics, as well as their associated infrastructural arrangements, that subtend such processes of extraction. Mediation is not simply nor always bound to technical objects. It is also a "living mediation."[7] Infrastructural mediation is a means to examine how colonial lifeworlds, sub-jectivities, and affects come into being through the design, building, mainte-nance, and repair of infrastructures that respond to resource frontier–making projects as settler media-making projects as well. It is to a subarctic resource frontier, a colonial environment where settler spaces and times are experien-tially made through the deployment of infrastructural arrangements, that I turn to ask: How are frontier-making projects themselves made?

Over the course of *Slow Disturbance* I become immersed in the world of the historical North Atlantic extractive fishery and remain submerged within a largely forgotten and overlooked microcosm of the British colonial domain: northern Newfoundland and Labrador. Britain's first colony was also its first resource frontier. I track this frontier through the lives of the settler fisherfolk who shored up the imperial fishery in this region from roughly the 1880s to the 1950s. In order to follow their fates across this frontier-in-the-making, I foreground how their lives and livelihoods were subject to forms of infrastructural mediation largely enacted by an evangelical Protestant medical mission established in the region to minister to these "toilers of the deep."[8] Medical doctor Wilfred Grenfell began traveling to the outports along the coasts of northern Newfoundland and Labrador in 1892 aboard the medical ship *Albert* sent by the Royal National Mission to Deep Sea Fishermen. In the popular North Atlantic imagination, Grenfell is a little-known and ambiguous figure: doctor, pseudo-saint, author, fundraiser, and missionary.[9] Yet in recent literature, Grenfell is seen to an ever-greater extent as a social reformer who "intervened to change the patterns of living" in northern Newfoundland and Labrador.[10] The mission he worked to establish, culminating in the incorporation of the International Grenfell Association (IGA) in 1914, would eventually oversee the construction and functioning of hospitals, nursing stations, schools, orphanages, cooperative stores, and light industries, among other institutional types, becoming a vast northern health network that the IGA ran until, in 1981, it was finally transferred to provincial control. Known as Grenfell Regional Health Services, it merged with the Health Labrador Corporation in 2005 to create the Labrador-Grenfell Regional Health Authority.

The Grenfell Mission is one of those submerged historiographic entities that is difficult though essential to call forth within the context of environmental and postcolonial media studies—below the surface of disciplinary attention and concern. To become immersed in its world is an attempt to open these fields, and media studies more broadly, to the marginal, so-called minor histories of mediation that inform how we can theorize the relationships between settler colonial projects and the emergence of mediating infrastructures.[11] For the mission was a forceful and environment-responsive mediating entity that relied on a host of infrastructures to reshape the resource frontier inhabited and created by the fisherfolk of northern Newfoundland and Labrador. Much like Tsing's ethnographic coproduction of the Meratus Mountains of Indonesia or the shaded and foraged matsutake forests spanning the globe,[12] this is an effort, shifted into the realm of media historiography, to think alongside the local livelihoods that gave shape to a particular resource frontier in the process of its becoming: "Frontier—not a place or even a process but an imaginative project capable of molding both places and processes."[13] The Grenfell Mission, informed by imperial legacies of social reform and colonial administration, devised an evolving set of infrastructural arrangements as the means of fashioning the right frontier. Each chapter that makes up this book pauses on how these infrastructures were bound up in the infrastructural mediation the mission was there to undertake—its evangelical and colonial imaginative project that projected a reformed resource frontier into the future through the building of a progressive and standard-setting medical infrastructure (chapter 1); the design of a radical and equitable system of cooperative finance (chapter 2); the mapping of the North Coast of Labrador through an experimental aerial surveying technique (chapter 3); and, finally, the mission's use of film to shape metropolitan perceptions of "need" prevailing on the coasts (chapter 4). These were attempts to shore up and recast the extractive conditions that prevailed in northern Newfoundland and Labrador from the late nineteenth century well into the twentieth.

Building on Lisa Parks and Nicole Starosielski's situating of a media infrastructure studies that can apprehend "the materialities of things, sites, people, and processes that locate media distribution within systems of power," *Slow Disturbance* extends this dynamic and relational approach to the formation of a historic resource frontier and its reliance on the deployment of infrastructure.[14] Here, as the case of the Grenfell Mission shows, this mediation was infrastructural in its refashioning of the fisherfolk's ambient environment. If, as Parks and Starosielksi contend, "infrastructures and environments dynamically mediate and remediate one another," it follows that this rescaling and investment in the minor, if deeply articulated, world of the Grenfell Mission and the fisherfolk of northern Newfoundland and Labrador can inform how scholars working across environmental, postcolonial, and media infrastructure studies theorize the site-specific forms of infrastructural mediation that sustain past and present colonial resource frontiers. As Ann Stoler notes, "'Minor' histories should not be mistaken for trivial ones. Nor are they iconic, mere microcosms of events played out elsewhere on a larger central stage." Rather, they suggest "a differential political temper and a critical space."[15] Cephalopod, extinction, extraction. These are all processes of infrastructural mediation that are sedimented within the Grenfell Mission as a settler infrastructural story, its hospitals, docks, airstrips, roads, cooperative financial systems, and aerial surveys constituting the infrastructural legacies that marked its North Atlantic field of operations. The mission's ongoing historiographical narrative that I both quiet and situate here is a means of mining how settler infrastructure making is an accretive process that binds together both the fisherfolk's lives as colonial pasts and the deeply felt settler colonial present. The very term *frontier* can sometimes obscure the real and affective homes that come to be built on the original displacements enacted by the promise of extraction. I deploy it here as a reminder of the temporal tensions that reside in settler colonial geographies between colonists (for whom frontiers become homes) and colonized (for whom frontiers are "the land").[16] This is a particularly Canadian legacy that permeates the settler state's claims to its territory. The settlers not only extracted, but stayed. As Patrick Wolfe has it, "Invasion is a structure not an event."[17] This ongoing legacy requires settler Canadians to denaturalize the environments they inhabit in order to allow their colonial past not only to become geographically legible but also to become infrastructural, that is, extended into the present across chains of settler colonial materiality and accountability.

I write this book in the hope of troubling the instrumental present-mindedness that underlies the staking and consolidation of resource frontiers through the current practices of extractive capitalism. The North Atlantic fishery, particularly its sad and telling manifestation in the province of Newfoundland and Labrador, is a surprisingly urgent and useful story of collapse.[18] The 1992 cod moratorium, a government edict that essentially marked the end of a five-hundred-year resource frontier, is a harsh object lesson in the ecological finiteness that actually underlies ocean resources. It was harsh in the toll it exacted on the people who were reliant on this frontier to sustain future-oriented livelihoods: "The activity of the frontier is to make human subjects as well as natural objects."[19] Resource frontiers, particularly oceanic ones, are indeed peopled. From forced displacements to colonial settlement, the ocean as a resource horizon is a case in point of an environment to be made available for extraction. Much like Macarena Gómez-Barris's exemplary effort to examine the subjugated if embodied knowledges that permeate contemporary Latin American "extractive zones," largely through creative Indigenous resistance movements, *Slow Disturbance* slides into a situated history of infrastructural mediation to uncover the lives and livelihoods of long-standing settler colonists in northern Newfoundland and Labrador.[20]

Like many colonial geographies, this settler colonialism is a sedimented overlay of waves of imperial arrival: stretching back to the sixteenth century, wave upon wave of largely French and British colonists took possession of the island of Newfoundland and parts of Labrador.[21] This historical colonial presence serves to reify and obscure the colonial identities of the fisherfolk that the Grenfell Mission was there to serve. By the late nineteenth century, the British imperial imaginary considered these "Vikings of Today" as an impoverished, exploited, and threatened population in need of aid; in some, if not all, respects, the colonizers had come to resemble the colonized.[22] The mission sought to hold up their white, Anglo-Saxon, and laboring bodies as particularly worthy of medical and other forms of imperial care, with Dr. Grenfell himself seeking to embody and act out a form of muscular Christianity capable of modeling the corporeal potential white fisherfolk bodies held.[23] Through immersion in the thick of this settler colonial society, this book sets out to map the under-examined divides between colonist (the fisherfolk) and colonizer (largely British mercantile firms engaged in the extractive fishery), with the mediating agency of the Grenfell Mission serving to mold this relationship through social reform–minded practical therapy. While I do not lose sight of the extinction of the Beothuk, nor of the progressive colonization and marginalization of Innu, Inuit, Mi'kmaq, and Métis across Newfoundland and Labrador, my aim is to articulate this historical resource frontier as a site of infrastructural mediation directed toward the fisherfolk as a group of settler colonists: How was this North Atlantic settler colonialism a mediating project? What sort of infrastructures did it rely on? What experiences of the resource frontier did the Grenfell Mission create? Taking up these questions requires readers to attend to the site specificities of extractive capitalism in this North Atlantic world, particularly as it was an apex of the British Empire's colonial environment making. In this context, the promise of extraction was made and performed by fisherfolk, colonial agents and administrators, merchants, and missionaries. Indigenous lives were co-opted by these extractive networks and affected by the Grenfell Mission's medical enterprise. They suffered from the promises of extraction. Whether "resource" frontier or "salvage" frontier, "where making, saving, and destroying resources are utterly mixed up," *Slow Disturbance* shows how extraction is predicated on forms of infrastructural mediation that were tailored, through missionary intervention, to the fisherfolk of northern Newfoundland and Labrador.[24] These historical and environmental echoes can be heard in current promises of extraction—sites of enclosure where commodities are made to emerge, sites of past and future fish.

Infrastructural Mediation

Resource frontiers are emergent. They become surveyed, staked, and extracted. Once a raw environmental phenomenon in a given location is exhausted, extractive practices shift to other sites. This book is immersed in the middle of the process of extraction—the *medium* of the resource frontier. Environmental media studies has begun to attend to the evolving set of material practices and substances that subtend the production and dissemination of media technologies, institutions, and discourses. Janet Walker and Nicole Starosielski specify that much of the scholarship coming under this banner, following that of Eva Horn, evinces "anti-ontological" approaches that shift away from essentialist conceptions of the constitution of media, whether institutional, discursive, or technological, and move toward ways of conceiving of media "as conditions of possibility for events and processes."[25] This instability underlying what media are and can be opens up the possibility of trying to account for the degree to which ecological conditions are intertwined with, or indeed constitutive of, mediating processes. Recent work by John Durham Peters has extended this environmental sensibility underpinning the medium concept to establish its interchangeability with environment making: "The old idea that media are environments can be flipped: environments are also media."[26] Peters establishes the existential stakes of expanded media, specifying how they provide "infrastructures and forms of life";[27] they are, following a tradition he situates extending back through Friedrich Kittler, James Carey, Lewis Mumford, Harold Innis, and Martin Heidegger, "modes of being."[28] I share Peters's call to uncover processes of mediation and the environmental media forms they rely on. However, rather than pursue his timely and important investment in non-anthropomorphic modes of communication, I shift this turn to environments as media in order to examine how the Grenfell Mission was part of a broader process of infrastructural mediation driven by a settler logic of control and a homiletic practice predicated on the materialization of physical good works—as I examine in chapter 1, the design, building, and maintenance of an equally colonial and infrastructural plant. This re-anthropomorphizing of Peters's environments-are-media argument is a means of rendering accountable the media-enabled existential conditions experienced by particular people across precise periods of time, notably at such acute sites where settler colonialism was moved along by waves of extractive capitalism. Resource frontiers make this tension between a given environment, mediating processes, and a laboring *anthropos* apparent; they are, like many media, always in between, relaying, caught in the making.

Infrastructural mediation also attempts to show how resource frontiers are reliant on modes of control that can account for and extend from their surrounding ecological conditions. This is inspired by the articulation of environmental media undertaken in Starosielski's *The Undersea Network*, where she notes how fiber-optic cable networks possess a complex interplay across distinct environmental contexts that oscillate between "strategies of insulation," which enable smooth connection regardless of their surrounding ecology, and "strategies of interconnection," which ground infrastructural arrangements in local social and ecological conditions.[29] The Grenfell Mission's particular form of infrastructural mediation was a hybrid of both strategies. For example, the medical infrastructure it built in the region was shaped by the length and habitation patterns along the coasts, which grew out of early forms of merchant philanthropy, while it also introduced entirely modern medical technologies, such as the x-ray. This infrastructural mediation attempted to account for the status the colony held as a multivalent dependency. Indeed, historically, the colony of Newfoundland was an integral part of the British Empire's network of resource-producing colonies. As one of the earliest comprehensive histories of the colony of Newfoundland remarks: "The island of Newfoundland has been considered, in all former times, as a great ship moored near the Banks during the fishing season, for the convenience of English fishermen. The Governor was considered the ship's captain, and those who were concerned in the fishery business as his crew, and subject to naval discipline while there, and expected to return to England when the season was over."[30] This was a near history that the mission had to contend with in devising its plans of medical and, eventually, social reform.

The *First Annual Report of the International Grenfell Association*, published in 1914, was a pamphlet intended for wide distribution across the mission's network, and it demonstrates the extent of the process of infrastructural mediation the missionary organization had accomplished since its beginnings in the 1890s. Ranging across its hospital ships, hospitals, and cottage hospitals (or nursing stations), the mission had treated 6,855 outpatients and 490 inpatients, with a total number of days of hospital care of 22,628.[31] Add to this the orphanage, school, and work of the industrial department in St. Anthony,[32] the headquarters of the mission on the Northern Peninsula of Newfoundland, as well as the King George V Seamen's Institute in St. John's,[33] and a more comprehensive picture of the scope of the infrastructural mediation the mission undertook comes into focus. Its physical infrastructure of care would be tallied, insured, renovated, built on, and generally improved on as the years passed (I touch on these efforts in chapter 1). New hospital boats would be built that could better withstand the ice and compensate for the poor knowledge of the subaquatic coastline. Yet running from roughly 1914 to 1927, while the mission moved toward a more centralized and bureaucratic system of organization that somewhat reduced Grenfell's idiosyncratic influence, the latter would nonetheless continue to be felt through a wide range of schemes of reform. Much like the cooperative system that he had put in place in the 1890s (chapter 2), which he continued to bolster through educational initiatives, Grenfell also brought to bear other forms of infrastructural mediation that drew on both the value placed on utility in the social gospel movement and Grenfell's faith that improvements could be made by directing social Darwinism toward the correct, Christ-centric and action-oriented steering influence. As a whole, these practices of infrastructural mediation make up a series of episodes in the mission's existence that foreground how its human subject of reform, while intensely local, was open to and integrated in multiple, mobile networks of global influence across such fields of endeavor as medical innovation, military production, and agricultural experimentation.

In *Slow Disturbance* I focus on four practices of reform that articulate the mission's understanding of infrastructural mediation that could minister to the fisherfolk of northern Newfoundland and Labrador. Their efforts at missionary reform espoused a particular version of the Protestant homiletic tradition that sought to shape both the lives of fisherfolk and their ambient environments through a series of infrastructural incursions. A more ambitious book could spend time examining the wider array of reformative practices, as I touch on below, that evince the mission's reliance on the building of infrastructural capacity with the aim of reconfiguring this North Atlantic resource frontier. A partial inventory of these reformative practices includes: the introduction of whole wheat flour into the Labrador household diet to combat malnutrition; the invention of a textile known as Grenfell cloth as a result of Grenfell's participation in World War I, during which he observed the need for a waterproof uniform for soldiers from fabric that took its origins in Labrador's fishing industry;[34] the implementation of a comprehensive local craft industry to produce hooked, silk-stocking mats and other goods that would be sold in London, Boston, and New York, with the entire enterprise coordinated by Jessie Luther, a pioneer in the occupational therapy movement; the introduction of rational dietary measures, largely based on research around the unlimited possibilities of the soybean by Dr. John Harvey Kellogg and, on the industrial aspect of its potential applications, the Ford Motor Company; the sustaining of an experimental agriculture division in contact with the American government's own installation at Rampart, Alaska, that also built numerous glass greenhouses with panes derived from a soybean compound; and finally, the use of publication and promotion, from magic lantern slides to travel books to Metropolitan Opera fundraisers, in order to create a donating public for the mission's philanthropic enterprise. While a constant behind all these practices of reform is the labor performed by the fisherfolk and the larger fate of the colony's fishing industry, the mission was working toward establishing new terms of mediation for northern Newfoundland and Labrador by shoring up the missionary organization as the sole medical provider to the settler fisherfolk. By situating an understanding of mediation that is inextricably shaped by its time and place, by its immersion in what Berlant calls the "lifeworld of structure," in this respect I follow other environmental media scholars who have taken up Sarah Kember and Joanna Zylinska's analogous argument that "mediation can be seen as another term for 'life,' being-in and emerging-with the world."[35] Parks and Starosielski describe how "this approach troubles any clear distinction between what we consider to be media infrastructure, such as a broadcast transmitter, and sites and processes typically thought of as its 'environment.'"[36] I would add that Kember and Zylinska's articulation of mediation opens up important stakes for media studies in examining the constitution of unconventional sites of mediation, such as resource frontiers, and how these sites are articulated in and through the organization of settler infrastructural projects.

These stakes largely revolve around a shift away from approaching media as discrete objects toward framing them within analyses that focus on them instead as "processes of mediation."[37] This argument has resurfaced across environmental media studies, media archeology, and other materialist analyses of mediating forms through their emphasis on the mobile constitution of "media" as not only "events and processes" but also, as Jennifer Gabrys's articulation of remote sensing technologies bears out, world-making infrastructures reliant on the durational aspects of mediation.[38] Kember and Zylinksa note in *Life after New Media* that mediation is a complex and heterogeneous process: the "originary process of media emergence, with media being seen as (ongoing) stabilizations of the media flow."[39] They work through what they see as the different incarnations of the term *mediation*. Ranging from Marxist theory's reconciliation of two opposing forces in a given society embodied within a given mediating object to mainstream media studies and its view of mediation as a "'mediating factor of a given culture' which takes the form of 'the medium of communication itself.'"[40] These, in their estimation, "structuralist" and "static" accounts of the process of mediation tend to focus on media "effects" brought about by distinct, identifiable, and usually exclusively human subjects that nonetheless work through and on material objects.[41] By way of contrast, for Kember and Zylinksa, mediation is an active and ongoing process of co-emergence at the biological and sociocultural levels ("that is an intrinsic condition of being-in, and becoming-with, the technological world").[42] The biopolitical sensibility that permeates their understanding of vital mediation was a powerful corrective to contemporaneous debates surrounding the emergence of new media as both scholarly objects and a subfield of media studies. This emphasis on the *lifeness* of mediation, and on media as temporal processes embedded in articulating contested conceptions of that vital experience, of being inevitably bound to these twin phenomena, establishes a conceptual bridge to my deployment of infrastructural mediation through this minor history of the Grenfell Mission. The mission story offers a prehistory of sorts to understanding how constitutive processes of mediation are to their social and environmental contexts—in this case, the fisherfolks' lives and biopolitical lifetimes—that the Grenfell Mission was so actively shaping through the practice of infrastructural mediation, that is, expanded, often ecologically inflected processes of mediation, from medical infrastructure to aerial surveying, that enabled and extended the operation of a particular resource frontier that registered these practices as an evolving environmental medium.

It is also an effort to shift recent articulations of "story" as a "material ordering practice," largely anchored in historical geography and its allied fields, into the purview of environmental media studies.[43] This book asks how settler infrastructural stories can be told in such a way as to highlight how they are bound up with processes of infrastructural mediation—that their ongoing and material durations have to be taken into account across anticolonial futures. It is by performing an infrastructural storying of the mission that I lay out a propositional methodology that can account for the *lifeness* of processes of mediation across such resource frontiers. While this method shares a sensibility with what Vivian Sobchack calls the "family features" of media archeology, most notably in its emphasis on the material dimensions of media as "forms and structures," it nonetheless shifts attention to the settler materialities and colonial capacities that tracing such mobile sites of infrastructural mediation present.[44] Rather than attempt an archeological excavation of the deep materiality of various media, whether narrowly circumscribed and technical or more broadly environmental, I pursue this method of infrastructural storying that asks analogous questions about how resource frontiers register processes of mediation that are durational and, in this instance, that extend settler colonial logics of extraction into the material order of the present. Infrastructural storying is a practice that sees stories as "modes of relation and intervention," a dimension of this method that I address in greater detail in chapter 4, and it asks environmental media studies to assemble stories about extractivism that can account for the sedimented processes of mediation that underlie its sense of legitimacy.[45] This sedimentation is not one-dimensionally archeological but rather extends across the practices that order how infrastructures are made, used, and repurposed—how both these practices and infrastructures come to co-constitute settler geographies of extraction. The Grenfell Mission as it appears in this book is thus one such manifestation of the practice of infrastructural storying. The mission story is one that works through the infrastructural dimensions of processes of mediation and how they can be made accountable for our current conjuncture defined by an extractive capitalism seeking out new and increasingly fragile ecologies; it is, indeed, a historical echo within the "extractive zone" that informs how contemporary media theory and environmental media studies can start to account for "medianatures" bound not only to a materialism of discrete media objects but also to a broader and more urgent set of oceanic, terrestrial, and atmospheric extractive practices: infrastructural mediations.[46]

Prehistories of Environmental Media: The Cod Fisheries

Across the province of Newfoundland and Labrador many of the buildings the mission built are still standing, the lie of the roads they traced paved smooth, and the localization of medical care in coastal communities throughout northern Newfoundland and Labrador an ongoing reminder of where the mission had been. While incorporated as the International Grenfell Association in 1914, "the Mission," as it is still referred to locally, was a vast, international network of volunteer labor. Women and men, as nurses, doctors, nutritionists, caregivers, teachers, craft instructors, carpenters, and bricklayers from all over Canada, the United Kingdom, the United States, and Australia, among other countries, came through the mission's headquarters in the town of St. Anthony and on to the mission's various nursing stations and hospitals along the coast of Labrador. All these laboring missionaries who were serving differing ideals of religious and secular obligation made up "a sort of Peace Corps."[47] This historical geography of missionary care belies the settler colonial reality underlying such a resilient and rooted town as St. Anthony. As with many settler geographies, the passing of time has concealed original forms of dispossession that supplanted one set of (Indigenous) lives for extraction-driven settlement—the fish were made to come first. This is a sedimented reality that stretches across the settler state of Canada, and it is made all the more apparent through many rural towns' thin veneer of aging colonial and "invasive" infrastructures.[48] These places show the signs of settlement. This history of dispossession requires deft negotiation given the sense of attachment to place that old colonies, such as Newfoundland, both celebrate and have a hard time acknowledging in relation to their absencing of Indigenous precedents—a colonial past "everywhere and nowhere at all."[49] *Slow Disturbance* ties together these histories by foregrounding how the infrastructural mediation the Grenfell Mission put to bear on the fisherfolk of northern Newfoundland and Labrador was not only part of this settler colonial project, but it functioned very practically to further marginalize Indigenous lives (and, later, land claims).

In many respects the mission operated at two speeds and across two distinct geographies. Its activities in Labrador came under the influence of its station at North West River, and mission workers there lived apart from, if alongside, Innu, Inuit, and Métis communities in the interior of Labrador and along its North Coast. As above all else a medical mission, providing essential services to parts of the colony that had only ever been served by intermittent medical cruises, usually once a summer, the mission engaged with Indigenous realities and made them a marginal and sporadic part of their reforming, colonial enterprise (with the Inuit residents of the North Coast of Labrador already under the long influence of Moravian missionaries stretching back to the seventeenth century). The Grenfell Mission's enactment of infrastructural mediation was both a settler colonial project and an everyday homiletic practice that would reshape fisherfolk lives, their North Atlantic extractive environment, and, in hindsight, relationships to the affective and practical management of a colonial order that persists today. This places equal emphasis on the emergent, durational character of infrastructural mediation that the book tracks across the Grenfell Mission story—it also makes my telling of the story into just such a media and historiographical event; like the promise of extraction, and as I will touch on in greater detail below, it is to be made and unmade through the Grenfell Mission archives, "generative substances" for epistemologies that pursue knowledge forms that privilege settler accountability,[50] and the lived locales across northern Newfoundland and Labrador that are a testament to "the way it was."[51]

This story is indeed a minor one.[52] Marginal. It is worth recalling that such a minor history of the mission can also really only come from a collection of partial impressions—divergent, spaced across time, and told from multiple gender, class, and institutional positions. One of the clearest indicators of the mission's human impact has been the memory work it has evoked.[53] Its myriad volunteers over the mission's nearly one hundred year existence were indeed marked by the experience. Often it was the pivotal moment in the arc of their lives. Mission work created a network of solidarity for many of its volunteers, with its fading, Victorian epitomizing of service continuously changing over the years. Yet for all its obscure characters and locales, it is nonetheless a radically situated story that builds on Peters's call for a turn to "infrastructuralism," the obscure, marginalized backdrops to both media theory as a field and the smooth apperance of modernity's westernized daily life (as Paul Edwards writes, "to be modern means to live within and by means of infrastructures").[54] Peters specifies that "infrastructuralism shares a classic concern of media theory: the call to make environments visible."[55] Resource frontiers, environments that highlight the ecological and extractive capitalist power dynamics that inhere in and by infrastructure, are often left out of the concerns of media theory. With notable exceptions that I will address below, it is as if the pioneering work of Harold Innis on the material dimensions of natural resource extraction and circulation has faded into and been absorbed by the more diffuse materialisms under examination in environmental media studies that couple specific media technologies to a suite of underlying energetic and resource-based "footprints."[56] Much like Parks and Starosielski's pursuit of a relational and materially interwoven media infrastructure studies, wherein environments are technologized and nonhuman entities not only possess agency but also constitute the ontological ground of mediation, this book sets out to highlight how processes of infrastructural mediation, such as those undertaken by the Grenfell Mission, can also be at the center of debates surrounding the constitution of contemporary extractive media environments.[57]

Early on in the emergence of environmental media studies, N. Katherine Hayles urged media scholars to apprehend how "nature" and "simulation" (largely articulated through virtual reality spaces) are not opposed but rather the result of a recursive flow of interpretation and experience: "Instead of accepting a construction that opposes nature to simulation, I seek to arrive at an understanding of nature and simulation that foregrounds connections between them. Not two separate worlds, one natural and one simulated, estranged from each other, but interfaces and permeable membranes through which the two flow and interpenetrate. Interactivity between the beholder and the world is the key."[58] Building on Hayles's bounding of real and projected environments, in a simliar vein Ursula Heise weighed the merits of "environment" as a metaphor in media theory. She argued that media ecology and its deployment of textual "environments" could benefit from grounding the metaphor in the political ecologies of spatial experience afforded by particular, real world sites.[59] In many respects scholarship across environmental media studies is an indirect response to Hayles's and Heise's attempts to account for the material and ecological substrata that undergird virtual media environments. This body of scholarship has far exceeded a simple green-ing of media ecology. It is infused by the political economic dimensions of Innis's work as well as Kittler's expansive definition of media as practices reliant on recording, storage, and processing and thus has taken shape around an investment in the politics and materialities of ecological situations and thought.[60]

This turn in environmental media studies could also be productively read as a return to the sited parochialism evident in Innis's early considerations of the railway, the fur trade, and the Atlantic cod fishery.[61] These sites of natural resource extraction, production, and dissemination could thus become the originary ground for the emergence of the ecological and material com-mitments of environmental media studies. In addition, through their em-phasis on, if not infrastructure in a narrow sense, then on infrastructure as a relationship-building phenonenon, as Susan Leigh Star and Karen Ruhleder observed,[62] these considerations of the material networks that build out from and across distinct resource frontiers prefigure a media infrastructure stud-ies capable of making its constituent environments visible and available for critique—as Liam Cole Young has it, Innis laid bare the "infrastructure of colonization."[63] Yet, as many scholars have noted, Innis's resource frontiers were not particularly peopled, with an impartial sort of accounting given to settler colonial practices and their effects on Indigenous communities.[64] As Peters generously puts it, Innis "was more interested in organization than in content."[65] *Slow Disturbance* takes up the task of tracing how the fisherfolk's extractive environments came to co-shape the Grenfell Mission's practices of infrastructural mediation. More than simply populating a particular resource frontier, the book thinks out from Innis's generative expansion of the mate-rial and ecological dimensions of the interrelations between extraction, infra-structure, environment, and staples (that often served the mediating func-tion of conventional media) with the aim of encouraging sited approaches to contemporary resource frontiers that take them as ecologically and politically defined by practices of infrastructural mediation that are capable of attend-ing to their attachments to histories of settler colonialism. The "extractive view" is one that is historically specific and assembled through material prac-tices.[66] As Innis notes in *The Cod Fisheries*, the North Atlantic fishing indus-try was reliant on a form of "exogenous" development. This was a constant, outward-looking mode of industrial and spatial organization. Through this structural arrangement, the coastlines of Newfoundland and Labrador, ex-tending up to just below the Arctic circle, became precarious transit zones, or *stages*, in a sense that goes beyond the term's designation of a utility-driven outpost for fishing; they were sites of mediated life. Having readers spend time on this existential and infrastructural stage is an essential hoped-for outcome of this book.

This topographic critique of Innis's work sometimes bypasses the "dirt research" that shored up his scholarship in political economy. Innis spent time on rivers, in forests, and traveling along with cod fishers on the North Atlantic.[67] It was fieldwork of a kind and one that relied on a firsthand apprehension of on-the-ground economic realities. Innis's writing on communication began in the early 1940s, with his book on the cod fisheries the direct predecessor to the dissemination of his better-known work on center-periphery relations and the formation of monopolies of knowledge reliant on space or time biases. For Innis, too, the fish came first, with the opening pages of *The Cod Fisheries* describing the physiology of cod, the composition of their eggs, and their richly specific marine environment off the Grand Banks of Newfoundland.[68] Published in 1940, the book sought to trace the gradual transition—or, in the case of the colony of Newfoundland, lack of transition—from forms of economic organization derived from commercialism to that of a capitalist system governed by economic growth, in the process offering a critical summary of the various pressures to which the Newfoundland fishery had been subject.[69] These ranged from the advent of machine industry to the substitution of the wooden sailing vessel with the steamship and the railway to the effects of improvements in refrigeration and consumer patterns in urban centers. In his chapter 14 ("Capitalism in Newfoundland, 1886–1936"), Innis sketches a biting diagnosis of the collection of forces, changing communication and transportation technologies foremost among them, that led to the end of responsible government in the colony. His treatment of how the relationship between social organization, mechanization, and technologically informed fishing techniques in Newfoundland shows how "Newfoundland was squeezed between two civilizations": "She produced for tropical countries with low standards of living, and had to compete with other foodstuffs and goods purchased from highly industrialized countries."[70] Unlike *The Bias of Communication* and its sweeping, patterned mode of historical analysis, *The Cod Fisheries* is an attempt to understand the local stakes of "the history of an international economy" across multiple generations of the colony's population. It was also a modulated work of advocacy, as Innis published the book in the midst of debates surrounding the projected independence of the colony.

The Cod Fisheries, with these contextual markers in mind, resonates more with current work in environmental media studies on processes of materialization, infrastructure, and global commodities than does Innis's more narrowly defined and acknowledged body of scholarship in media theory. "Like more recent theorists," Jody Berland writes, "Innis viewed colonial space as traversed space; not the empty landscape of a wilderness, or geometrical, abstractly quantifiable space, but space that has been mapped and shaped by specific imperial forms of knowledge and administration."[71] Innis's geographically and historically multilayered portrait of the fate of the colonial cod fishery is a reminder that each resource frontier is nonetheless a "foreign form requiring translation," a spatial and temporal colonial project of infrastructural mediation.[72] Slow Disturbance shares Innis's investment in troubling the subjectivities and infrastructures that stem from cod: fisherfolks and missionaries, hospitals and aerial surveying.

Yet what to do with Innis today? How to think alongside his insights on colonial resource practices rather than merely place his influence in a genealogy of media theory? After all, resource frontiers, as Tsing claims, rely on "traveling theory."[73] This reminder of the fish coming first, prior to Innis's investment in untangling imperial webs of communication, situates the rural periphery as a material site for thinking through the stakes of staples always destined to travel. Whether grain elevators on the Canadian prairies or coastal cable-landing sites in Hawaii, the full spectrum of Innis's scholarship on colonial topographies foregrounds how the often forgotten rural is the de facto ground of networks of trade, transporation, and communication.[74] One aim of Slow Disturbance is to provide dimensionality and depth, equally historiographic in range, postcolonial in perspective, and topographical in scope, to the marginalized transit zones of historical extractive capitalism. For the Grenfell Mission, northern Newfoundland and Labrador, sites of both extraction and transit, held the promise of a new colonial order. Much like Peter van Wyck's exemplary mining of Innis's "territorial archive" across the "highway of the atom," this book follows the fate of the fisherfolk of northern Newfoundland and Labrador as shaped by the Grenfell Mission's efforts at reform. This is an infrastructural trail that leads at once back in time to the colonial fishery and into an unknown future around North Atlantic extractive frontier making.[75] I take inspiration from Innis's emphasis on the structural, political, and economic conditions of resource frontiers, and I aim to follow this line of inquiry throughout in order to examine the experiential modalities that underpin resource frontiers, that is, how the Grenfell Mission relied on particular processes of infrastructural mediation to shape the subjectivities of these North Atlantic settler colonists.

Settler Infrastructure

Over the course of the 1880s, settler colonists engaged in the fisheries, especially those on the coast of Labrador, experienced several failed seasons. Newfoundland was a colony reliant on a single resource, and when this resource failed to deliver the needed standard of living to its labor force, the effects of this failure circulated widely and quickly, worsening what were already precarious subsistence-living conditions in much of the colony. In 1891, the premier of Newfoundland, Sir William Whiteway, led a delegation to London to protest the imperial government's imposition of restrictive fishing policies on the island's west coast. Whiteway, in a report addressing the possibility of building a railway in Newfoundland, pointed to the structural economic conditions that were leading to its financial and social precarity.

> The question of the future of our growing population has, for some time, engaged the earnest attention of all thoughtful men in this country, and has been the subject of serious solicitude. The fisheries being our main source, and to a large extent the only dependence of the people, those periodic partial failures which are incident to such pursuits continue to be attended with recurring visitations of pauperism, and there seems no remedy to be found for this condition of things but that which may lie in varied and extensive pursuits.[76]

This open-ended call to action was picked up by Francis Hopwood, a council member of the Royal National Mission to Deep Sea Fishermen and an assistant solicitor at the Board of Trade. On the authorization of the fishermen mission's council, Hopwood traveled to Newfoundland in the fall of 1891 to assess the possibility of bringing the fishermen's mission's activities to Britain's oldest colony.

Hopwood's visit did not take him to the remotest regions of the colony, yet through the secondhand accounts of such public officials as judges and government adiminstrators, as well as the private interests of newspaper editors, merchants, and clergymen, his findings ranged across the social, political, and economic conditions of those engaged in Britain's broad network of migratory fishing. With the decline in export prices for saltfish in the 1880s; the increasing protectionism of the French market; the ubiquitous practice of the truck system, which kept fishermen in a state of constant indenture to local merchants and functioned on a barter system that excluded cash, thus forestalling the accumulation of capital savings; and, finally, the near absence of medical care, basic forms of administration, and law-keeping for the close to twenty-five thousand "floaters" who made the trip every summer to fish off the coast

FIGURE I.4. "St. John's immediately after the fire." Photograph from MS 254, Wilfred Grenfell's personal album, 1889–1892, Wilfred Thomason Grenfell Papers.

of Labrador, all these factors combined to spur Hopwood to recommend in his report that the fishermen's mission send a hospital ship the following summer. While Hopwood was careful to specify that this was to be an "experiment" rather than a permanent "institution," he was nonetheless surprised at the conditions to be found among this settler Anglo-Saxon population and was confident that regulation would improve their conditions over time.[77]

On the council's recommendation, the *Albert* was to set sail for the colony of Newfoundland from Yarmouth, a small port town in Norfolk, England, on June 15, 1892, with Dr. Wilfred Grenfell commissioned to serve as the hospital ship's physician. Upon their July arrival in St. John's, they found mostly charred ruins, the city having been virtually consumed by fire just a few days before. With almost every doctor in town unable to take patients, Grenfell established a temporary clinic onboard the *Albert*. With an eye on their departure for the Labrador coast on August 2, Grenfell had already become acquainted with a situation of scarcity and need and a disturbing, blackened tabula rasa for his emergent missionary practices and spiritual sense of action. "I always have the feeling that, if we are to do a missionary work with the spirit which I feel alone is of any value," Grenfell writes, "our Mission should realise from the beginning that it is a work of sacrifice."[78]

THE GRENFELL MISSION CAME at a time of infrastructure. The Anglo-Saxon fisherfolk they were there to minister to were in need of a form of infrastructural care that was both hard and soft: hospital beds and cooperative finance. It was a form of care that brought together the mission's emphasis on practical therapy and its turn to infrastructure building and maintenance in order to create what they saw as reformed social, economic, and, in time, environmental conditions for the fisherfolk to flourish under the broad tenets of the social gospel movement. Their North Atlantic resource frontier was just that, a frontier that had to be settled and resettled, accommodated, and brought up to a mobile imperial standard that was enacted periodically and haphazardly, largely depending on a given colony's strategic importance in the British Empire's wider network of trade. Grenfell, pursuing a brand of muscular Christianity driven by practical action, was ready to deploy a full spectrum of reforms to reshape not only the economic conditions that prevailed on the coasts but also the very social and environmental realities that he found upon his arrival in 1892—a meeting of indifferent ocean, wind-contoured coasts, and indebted settler colonists who fished to go on living. The Grenfell Mission was an infrastructure-making project—and, so, a practice of perpetuating and projecting settler lives.

The array of reformative practices listed above were part of a responsive, if not particularly coordinated, strategy on the part of the mission to fashion what they saw as a more equitable, racially sound, and spiritually driven colony that would work in symbiotic extractive harmony with its North Atlantic ecology.[79] The fisherfolk were indeed "toilers of the deep," and the "deep" was an environmental zone wherein laboring men could acquire hearty morals and physiques through extended and intimate exposure to the hardships of cold, dark North Atlantic waters. The mission had this fascination with and commitment to infrastructuralism: how building, meeting standards, and devising practices of reform and care could shore up the colony. They were builders of colonial infrastructures that had to stand in for imperial forms of administration and a robust "state of the nation."[80] The mission was invested in the propagation of an evangelical Protestant logistics: a homiletics that took infrastructural development as its central tenet. "The job of logistical media is to organize and orient, to arrange people and property, often into grids. They both coordinate and subordinate, arranging relationships among people and things."[81] North Atlantic things (including people) could be improved. Thus, the mission's evangelical reformative practices were infrastructural in their attempts at constantly remaking and reforming the relationships between the fisherfolk and their ambient, extraction-driven environment.

This merging of reformative practices and sited environment starts to complicate how the story of infrastructure, particularly in colonial contexts, is "one of disconnection, containment, and dispossession."[82] The Grenfell Mission's role in building, maintaining, updating, and repairing a settler infrastructure is an untold story as well as a contribution to foregrounding how site-specific histories underlie faded, ruinous, and just simply paved-over settler infrastructures. As Deborah Cowen asserts, "Infrastructures reach across time, building uneven relations of the past into the future, cementing their persistence. In colonial and settler colonial contexts, infrastructure is often the means of dispossession, and the material force that implants colonial economies and socialities. Infrastructures thus highlight the issue of competing and overlapping jurisdiction—matters of both time and space."[83] The durational dimensions of infrastructural mediation can account for how the mission's infrastructure making coalesced across decades, and this process was indeed reliant on means of dispossession in order to assemble a reconceived settler colonial resource frontier. *Slow Disturbance* situates the mission's ministry within an infrastructural zone that aimed to bring the settler colonist fisherfolk up to international medical, financial, and cartographic standards.[84] Each chapter examines how such an infrastructural zone came into being through the incremental reduction of infrastructural difference the mission undertook through its elaborate practices of reform—infrastructural difference could be made to account for the material lack that the "neglected" settler fisherfolk experienced. "Infrastructure is by definition future oriented," Cowen writes; "it is assembled in the service of worlds to come. Infrastructure demands a focus on what underpins and enables formations of power and the material organization of everyday life."[85] The Grenfell Mission staked its claim in remaking and projecting this particular North Atlantic colonial world. Their infrastructural work was a case of both imagining alternative infrastructures to those of the prevailing and usually inequitable imperial and capitalist resource economies while also making that work complicit in the maintenance of a settler colonial project.

This specificity gives the minor Grenfell Mission story the capacity to shift current conceptions of settler colonial infrastructure toward a nested and almost circular history that oscillates between repair and maintenance stemming from an original, if drawn out, settlement of dispossession. The settler colonial project itself becomes a metaphoric road that has to be constantly maintained, filled in, touched up, and tweaked in going from dirt to gravel to asphalt. The Grenfell Mission was precisely engaged in service to the building of *multiple* worlds to come, equally material as spiritual, which makes of their infrastructural work itself a part of its homiletic practice. As the chapters that follow bear out, infrastructure for the mission was the "process of relationship building and maintenance" of both material standards of improvement and the modes of existing within those standards, with both of these substantively modified by the prevailing ecological conditions of northern Newfoundland and Labrador.[86]

As "The Way It Was, St. Anthony, 1959" will touch on in more detail, it is the Grenfell Mission's infrastructural legacy that stretches on most definitively into a future that is still in the making. This processual and time-based conception of mediation is also one that can extend to phenomena, such as infrastructure, that are process-based, future-oriented, and encompass human and nonhuman agencies.[87] Attending to the forms of infrastructural mediation that produce them across resource frontiers can allow for an apprehension of their becoming, that they are relationships which emerge, that are durational. In the early days of the twentieth century, the Grenfell Mission intuited that it was through infrastructure building that a reformed colonial reality could come into being. Infrastructure was the impetus and product of an imperial resource frontier's real and imagined geographies. In a settler state such as Canada, and as Innis delineated in his tracing of the resource routes drawn by the fur trade and other extraction-driven forms of colonial commodity production, infrastructure would become concretized as a phenomenon that is at once ecological and social, material and relational—as Berland so aptly puts it: "The wheat cannot be understood separately from the train, and vice-versa."[88] So too, today, the promise of fish in a place such as St. Anthony, the mission's former headquarters, cannot be understood separately from the infrastructural work that the Grenfell Mission put in over the course of much of the twentieth century. What lies both behind and beyond this past promise of fish and, more specifically, the promises of North Atlantic extractions to come? The answer lies outside of the realm of the cold-water shrimp fishery that has gained momentum in northern Newfoundland over the past decade or so. Behind this commitment to extraction lies the imaginative capacity that inheres in infrastructure building. Alternatives to prevailing conditions emerge when infrastructure is built. This was a foundational principle that the Grenfell Mission articulated through their evangelical lens—maintenance and repair of infrastructure meant the maintenance and repair of fisherfolk lives. (I address this in more detail in chapter 2.)

This commitment to a conception of infrastructure that is hard as well as soft, material and social, echoes current investments in staking a sense of hope in the alternative politics (and lifeworlds) certain infrastructural commitments can enact. I extend and articulate Berlant's understanding of "living mediation," which parallels Kember and Zylinska's understanding of the process, in order to draw attention to how infrastructure can move along with and reflect a given social formation's relational becoming.[89] Infrastructural mediation is suggestive of potential re-form as well. This echoes the Grenfell Mission's mobile conceptions of infrastructural care and work that could sustain the emergence of alternative economic and social conditions for the settler fisherfolk. It is the very longevity of the mission, with a presence in the colony, then province, from the 1890s until the 1980s, that provides a slowed down, colonial lens through which to apprehend the making of a settler infrastructural world: "What constitutes infrastructure in contrast are the patterns, habits, norms, and scenes of assemblage and use. Collective affect gets attached to it too, to the sense of its inventiveness and promise of dynamic reciprocity."[90] How to track across this processual reciprocity? How to make out, through immersion in the formative moments of these practices of infrastructural design, maintenance, and repair, infrastructures as mediating processes?

It is also here that the mission's infrastructural work exceeds its missionary institutionalizing frame. Grenfell often characterized the mission's work as self-eliminating. In time, the fisherfolk would be fully independent, healthy, faith-abiding, and productive settler colonists once more. In the final issue of the Mission journal, *Among the Deep Sea Fishers*, which appeared in July 1981, Dr. Peter J. Roberts, then executive director of Grenfell Regional Health Services, opens the issue with his impressions of the "process of change" for the mission that was coming to an end.[91] Roberts uses the trope of a single trip on the mission plane, one of innumerable routine flights he took on the Northern Peninsula, from Roddickton to Deep Harbour, to review the IGA's past in what he admits is selective and sped-up recall. "One cannot consider life in this area without knowing 'the Mission,'" he writes. "Undoubtedly, there was life here before the Mission and there will be life here after it is gone, but as long as it existed the IGA was an essential part of the life of Northern Newfoundland and Labrador."[92] In an echo of Grenfell's own moral purposefulness, Roberts sets out the essential, enduring core of the mission: "This trip through time clearly isolates the essential fact that people have served with, and for, their fellows and that no matter how grand or menial their work may have been they have contributed to this worldly life. No mere detail must obscure this fact for herein lies the greatest achievement: the Mission provided the means for all these people to serve their fellow man."[93] This is the above all else infrastructural collective affect that lives on from the Grenfell Mission's enduring commitment to infrastructural mediation. It was and is an alternative material and social infrastructure that has attained a regional reality in northern Newfoundland and Labrador, a settler infrastructure of repair and renewal that has settled into the current real and imagined boundaries of its North Atlantic resource frontier. Environmental media studies, and the environmental humanities more broadly, can begin to attend to the submerged affective registers that are shared, shaped, and coextensive between human and nonhuman agents and that are set into relation by infrastructural arrangements.

The case of St. Anthony, as a sited placeholder for the mission's regional influence as a whole, exemplifies the tendency to overlook mediation as a "process of change," as Roberts would have it, which takes place between real world institutions, people, and diverse human and nonhuman infrastructures. Mediation is not only oriented toward a movement of channeling and becoming; it is also a process capable of registration and stasis. Mediation can be cyclical and take place in situ. Today, St. Anthony, with its deep harbor and a long relationship to the Atlantic at its mouth, has a fifty thousand square foot cold storage facility, factory-freezer trawlers that sit at its edge, and an impressive communications antenna atop the rise that marks its North Atlantic entrance. It has a shrimp-processing facility, jointly owned with Clearwater Seafoods, that processes roughly five and a half million pounds of cold-water shrimp per year.[94] It has a tourist trade built up around the Grenfell Interpretation Centre and the whales, icebergs, and majestic scenery that are a short boat ride away. It has the Charles S. Curtis Memorial Hospital, an institution that serves the Northern Peninsula and Labrador for a range of specialized medical services. It has the Viking Mall, St. Anthony Elementary School, Harriot Curtis Collegiate, and the Polar Centre, which comprises a hockey arena, conference center, and an indoor walking track. St. Anthony has roads, streetlights, a traffic light, a sewer system. Within the province cum colony's history of remote, poverty-stricken outport communities, St. Anthony would seem to have it all. Yet what it lacks is an open-ended and secure sense of a future. As with many industries in the province, St. Anthony's future is seasonal. A looming threat is the onset of a prolonged economic winter. To a serious degree, these are problems of infrastructural mediation, that is, a resource-driven societal conjuncture defined by its infrastructural networks of exchange, time, and distance. What the town does nonetheless possess is a profoundly anchoring past that is shaped by the affective ties the mission's infrastructural care and work performed over so many decades. It is a place of settled infrastructure. While no longer the missionary outpost of old, St. Anthony is sustained through its own performance of the minor and marginal Grenfell Mission story. Its residents, particularly those who experienced the heyday of the mission's influence in the 1950s, can attest to the "inventiveness and promise," to the "dynamic reciprocity" that inhered in what the mission built, arranged, made happen—an affective settler infrastructure that lives on in all manner of community groups, voluntarism, town festivals, and a diffuse pride in "the way it was" made.[95] "Alternative worlds require alternative infrastructures," Cowen writes, "systems that allow for sustenance and reproduction."[96] Resource frontiers give rise to a horizon of infrastructural worlds—settler, colonial, and settled.

FIGURE I.5. Signage for the Viking Mall in St. Anthony, Newfoundland and Labrador, 2011. Photograph by author.

Francis and Agnes Patey's living room window looks out at the mouth of St. Anthony's harbor. During the right season, what fishing boats that are left are bound to pass into view. Whales occasionally make an appearance. Near the end of Lamage Point Road, their mint green house stands sturdily and neatly among the hard and receding rocks that make up the ground in these parts. Across the water are Fishing Point Park and the United Church cemetery. From this distance the cemetery resembles a pasture in miniature, what with its white fencing and gravestones made out as oddly shaped cows.

Francis Patey showed me his workshop when I first came in. Located in the basement of the house, the work area is tucked away beneath low rafters that keep you alert, even though plenty of soft foam is wrapped around them. Looking ahead to the Come Home Year festival in town, that would take place in July 2012, he was in the process of making some crafts that document the years gone by. Made of wood, paint, and Mr. Patey's own painstaking efforts, his models re-create the objects that went into constituting the inshore fishery. Small stages of wood. Boats, loaded to the gills with fish, sitting deeply in the water and signaling a good catch to onlookers on the shore. Mr. Patey has looked back to the way it was so as not to forget. He is the author of seven books, excluding an unpublished collection of hate mail and press clippings from the 1960s surrounding the debates around the seal hunt, which Mr. Patey keeps in a large green scrapbook.[1] His small-scale objects are what is left of a fishery that has been taken over by industrial concerns and trawlers. He's not nostalgic over the years of struggle and hardship that he and his family endured—he knows that so-called progress comes at many costs, some unknowable. Times change. Yet Mr. Patey spends hour after hour producing his models of a fringe reality that younger generations deem as distant labor. The fishermen of the Northern Peninsula may as well be Greek gods: an oilskin-clad sealer caught in the midst of an epic clash with the indomitable Bardot (see "The Way It Was, St. Anthony, 2011").

Agnes Patey was, for a very long time, the president of the local Grenfell Historical Society. When asked why she put so much time and effort into her work, she responded that, for her, the hospital was her home. Afflicted by a tubercular spine at the age of two, she spent her childhood in the local hospital. It was where she lived. For her, hospitals are important places. They are not just for patients. Hospitals are also representations of the life of a community. As the hospital goes, so goes St. Anthony. That's the way it is.

———————

The fish came first. That's the way it was. How to make the past present here? *"Frontiers aren't just discovered at the edge, they are projects in making geographical and temporal experience."*[2] This book assembles its own experiential horizon of the Grenfell Mission story, a documentary enterprise largely grounded in the mission's archival present but that also proceeds through a method that attempts to follow such cases of settled infrastructure back through their lived processes of mediation. If, as Deborah Cowen claims, infrastructures are largely suggestive of futures, what pasts inhere in their mundane appearance today? With just such a question in mind, over the course of 2011 and 2012 I traveled to many of the principal sites that made up the Grenfell Mission's infrastructural network. Through formal interviews and informal conversations with municipal officials, teachers, doctors, fisher people, and all manner of local residents at those socially magnetic Tim Hortons across northern Newfoundland and Labrador, I undertook somewhat haphazard fieldwork of a kind that was, at first, merely after a sense of the temporal markers of the Grenfell Mission legacy: How were places like St. Anthony and the people who lived there part of an experiential dimension of the mission's enduring influence? How did the mission live on in social practices that sustained life in these particular communities? How was here *here* and not somewhere else? This led me toward a more sustained and systematic, if still experimental, effort at a thick description of infrastructural appearance.

"THE WAY IT WAS," as an eighteen-point phrase glued onto his models with care by Francis Patey, is a marker of the embodied gulf that separates zones of experience on resource frontiers—how fisherfolk labored from skiffs to trawlers. In this book, "The Way It Was" serves several interrelated purposes. First, it allows for the voices of some of these local residents, including Agnes and Francis Patey, to articulate how they experienced the mission. Interspersed over the course of the book, these sections allow for readers to grasp firsthand experiences of the mission's settler affective infrastructure.

How largely, if not exclusively, now-elderly residents of northern Newfoundland and Labrador lived as part of the mission's sphere of influence. This not only complements the chapters' focus on the period between roughly 1890 and 1950, but it also echoes how infrastructural mediation is a process reliant on the experiential dimensions of duration: How did places like St. Anthony or North West River experience the mission's understanding of change? These sections, in giving space to firsthand reportage of a kind, follow the broader documentary logic of the book as a whole—what I describe below as a form of "slow historiography" with relevance to environmental media studies, media infrastructure studies, and media history in its emphasis on tracking the durational dimensions that inhere between mediation and settler infrastructure building. I will outline in greater detail how I see this historiographical gesture opening up possibilities for thinking through the relational emergence between affective settler infrastructure and archival work. I pair this turn to historiography with two analogous endeavors: the experimental fieldwork I briefly sketched in the introduction and, connected to this effort, a visual method that attempts to draw the book together as an experiential horizon of the process of infrastructural mediation itself. In what follows I address all three of these interconnected dimensions.

While I spent hour after hour in a host of archives that collectively housed the records of the Grenfell Mission as a missionary enterprise, it was the maps the mission produced that called me out into its world. Maps of the North Coast of Labrador. Quickly hand-sketched maps used to facilitate navigation to seldom-visited bays. Minute and detailed maps that displayed tubercular rates in the colony. St. Anthony, Red Bay, North West River. There was, again following Peter van Wyck, this "territorial archive" that demanded careful attention. After all, these were the sites where the mission's infrastructural legacy had led—old and new, roads and docks, leaving and living territorial traces mirrored in a missionary past and municipal present. In tracing how the Grenfell Mission put to work varied practices of reform in the service of their project of infrastructural mediation, of reshaping this particular colonial resource frontier, this called for a broad field of inquiry—fieldwork, archival practices, visual methods—that could attend to questions of duration, affect, infrastructure, and the settler colonial project.

In this respect, *Slow Disturbance* is in conversation with recent efforts in media infrastructure studies to genuinely follow the relational emergence that infrastructure building gives rise to. In broad terms, this is emblematic of a now long-standing reengagement with topography and, more specifically and currently, a political ecology of places that are rich with human and nonhuman experience. Whether "network archeology"[3] or "territorial archive," scholars across media studies are devising the means to document an expanded field of media-related sites, practices, and stories. This stems from infrastructure studies' now well-worn maxim that the infrastructural condition "is something that emerges for people in practice" and is "connected to activities and structures."[4] Turning back to document how this relational becoming occurred in particular locations, asking about what those activities and structures resembled and who made them—these are the questions that open settled infrastructure up to a fieldwork that pursues the experiential horizons of a particular case of infrastructural mediation. A territorial archive, certainly, but also an experience of that archive's organization of the ongoing social "dynamic reciprocity" that infrastructure facilitates, according to Berlant—the territorial archive, "dispersed, but in situ."[5] Figure and ground, object of study and material context, here become intertwined. Media environments, to circle back to Heise, are materializations of sited ecologies, spatial practices, and sedimented human experience. *Slow Disturbance* extends this emergent lineage by bringing an indeterminate ethnographic sensibility to bear on the book's wider collection of methods that all seek to document this figure-ground relationship: how the promise of extraction contains a geography of temporal and spatial experience that, for the Grenfell Mission, constituted its project of infrastructural mediation. This book is an effort to map those horizons of experience that do not so much run forward from the past into the future, but are instead concretized in the Grenfell Mission story that today circulates through places such as St. Anthony—that are in situ as hospital, road, and dock.

This leads to the underlying question that I have thus far left aside: What do all these pictures want? Or, what do I want you to want from these pictures?[6] Over the past number of years various critiques have circulated pointing out a visual bias in infrastructure studies.[7] This book's investment in documenting the experiential horizons of infrastructural mediation is certainly limited to its chosen documentary media: writing and photography. The latter works out from the relational commitments that stem from studying what could be thought of as the topographies of infrastructural emergence. In this book I deploy photographs across two registers. The first register seeks to document where and under what conditions this process of infrastructure emergence is taking place. These photographs can all be read as contemporary markers of the Grenfell Mission's settler infrastructural legacy. The second register contains photographs of a broader array of archival media—letters, advertisements, maps, magic lantern slides, historical photographs, and more.

Maps depicting the extent of the Grenfell Mission in 1906. From *Among the Deep Sea Fishers* 3, no. 4 (January 1906).

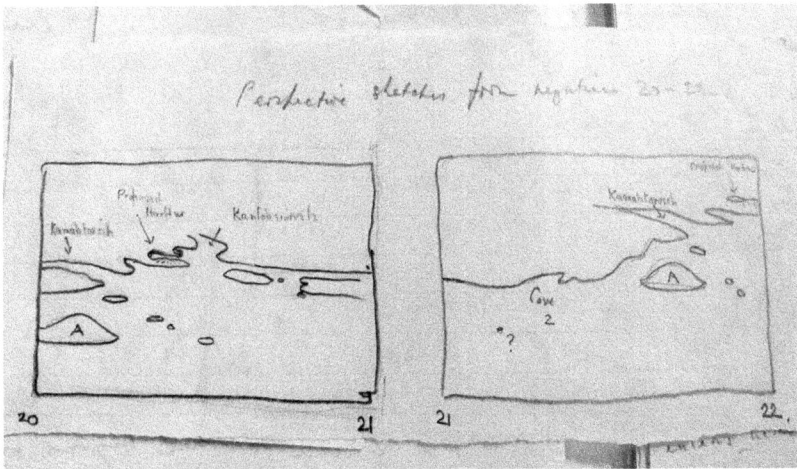

"Perspective sketches from negatives 20–22." From the Forbes-Grenfell Survey of Northernmost Labrador, 1931–35, Alexander Forbes Papers, Francis A. Countway Library of Medicine, Center for the History of Medicine, Harvard Medical Library.

Map tracing of northern Labrador from the Forbes-Grenfell Survey of Northernmost Labrador, 1931–35, Alexander Forbes Papers, Francis A. Countway Library of Medicine, Center for the History of Medicine, Harvard Medical Library.

Following this documentary path throughout the book could be thought of as following the open-ended infrastructural mediation enacted by the mission. Beyond the territorial archive itself, these documents are all that remain of the mission's incredibly comprehensive efforts at reshaping the lives of the fisherfolk; they are salty documents in a way, produced in and by missionaries along the coasts of northern Newfoundland and Labrador as well as farther afield in metropolitan offices in New York and London. Both of these registers are interspersed throughout the chapters and should of course be read in relation to the stories I tell therein. Affect in voice. Territory in image. "Generative substances" in documents. Storying settler infrastructure. Such are the methodological propositions this book makes.

Across *Slow Disturbance* these registers blend on occasion and foreground the material, processual constitution of photography as a practice of image making. "Slow Disturbance, '5 Canada,'" for instance, is a photograph of a framed stamp sheet that hangs in a corridor of the Charles S. Curtis Memorial Hospital in St. Anthony. While the commemorative stamp holds evident interest in its representation of Grenfell piloting a hospital schooner, as well as in its denoting the missionary's place in the Canadian historical imaginary, I deploy it here as a reminder of the iterative intrusion of mediation as a process of co-constitution between media of capture, such as photography, and anchored worldmaking sites, such as St. Anthony. Throughout the book, each image is indeed a photograph rather than an archival reproduction, as each is an index of the slow disturbance mediation performed by historiography, archives, and fieldwork.

As I address in greater detail in chapter 4, slow disturbance mediation is an effort to adopt ecology's capacity to foreground processes of change that are grounded in vital and symbiotic conceptions of human and nonhuman life.[8] It is a reminder that the forms of mediation registered by an environmentally inflected medium such as a resource frontier are accretive, waves of disturbance that bind together settler and shore. The structure of the book intentionally figures mediation as a hermeneutic experience in order to highlight how this medium too possesses its own infrastructural logics of sense making. In a similar vein, "Slow Disturbance, 'Channel 12'" documents the social ties broadcast by St. Anthony's community channel and constitutes another instance of mediated repetition: the boxy television containing bands of light that pulse across the screen and the announcements that establish long echoes of voluntarism and local philanthropy as well as a very specific sense of duration and durability that is part of this infrastructure of community communication. Like the case of some Grenfell Mission footage that I treat in chapter 4, more specifically its capacity to act as what I call indeterminate media that foreground processes of mediation and their ties to the durational dimensions of particular environments, my photographic practice throughout the book serves as an indexical anchor in the processual infrastructural mediation that sites such as St. Anthony seem to demand in their open-ended unfolding. Living mediation is, after all, an ongoing process that can only be stilled, captured, momentarily paused in the service of making it tell a story—putting materials into an order, making this ordering a practice.

Yet what of the mission's archival agencies that demand their own forms of attention? Figure and ground. Document and archive. Ann Stoler's characterization of documents as "generative substances" for historians grappling with the Dutch colonial enterprise is not only a useful reminder of the material substrate that comprise archives, but it is also a destabilizing of the document's relationship between its presumed materiality (usually paper) and its semantic and interpretive possibilities—documents have "itineraries of their own."[9] Indeed, documents, "unlike information . . . are importantly situated; they are tied to specific settings."[10] The Grenfell Mission archive is a collection of such productive documentary materials, and I deploy them here as such generative substances, raw phenomena of a kind that can find a place within the resource frontier–making logic at work across the colony's North Atlantic outports. If, as Esther Leslie claims in relation to the extraction and transformation of coal during the Industrial Revolution, materials are "transformative, transitory, non-eternal, productive," then how can the material properties of documents be foregrounded in order to highlight their interpretive malleability? Is it "possible to tell history from the standpoint of matter— coal, diamonds, gold, metals, glass, dyes, cellophane"?[11] Is it possible to tell history from the standpoint of living mediation? Following this historiographical logic entails treating the Grenfell Mission as a case of infrastructural emergence and allowing readers to assemble these materials into their own hermeneutic patterns of matter-meaning.

It is this gradual, accretive process that suggests a historiographical parallel with Rob Nixon's tracking of the unspooling consequences of a slowed down and cumulative understanding of the agency of detrimental environmental effects. Current appropriations of Nixon's concept somewhat neglect the methodological orientation at the center of his argument: "We also need to engage the representational, narrative, and strategic challenges posed by the relative invisibility of slow violence."[12] Resource frontier making is an experiential project, with its temporal dimensions encompassing the representational stakes of trying to account for the histories it gives rise to. On such frontiers like St. Anthony, this project's ecological effects are indistinguishable from its arrangement of an infrastructural disposition—the geographic and temporal experience of extraction becomes "the living mediation of what organizes life."[13] As Nixon asks: "How can we convert into image and narrative the disasters that are slow moving and long in the making, disasters that are anonymous and that star nobody, disasters that are attritional and of indifferent interest to the sensation-driven technologies of our image-world?"[14] This book pivots around a resource frontier's process of infrastructural mediation. It is an effort to document and display the durational dimensions of this mediation as they are made manifest across both territorial and material archives. Environmental violence "needs to be seen—and deeply considered—as a contest not only over space, or bodies, or labor, or resources, but also over time."[15] Thus my emphasis on *Slow Disturbance* both embodying and giving rise to experiential horizons of resource frontier making, of settler infrastructure as it is and was that becomes arranged relationally with its own stories of emergence.[16] This makes of the book an instance of "living mediation," an experiential arrangement of visual and textual historical substances, aural accounting, and a general anti-representational sensibility turned toward unsettling the making of resource frontiers. The way it was. The way it may come to be.

North West River arrives at the dead end of a sixty-kilometer road. Relics of the mission, largely wood-frame houses with green trim, populate its grid of streets. Up until the late 1970s no bridge connected North West River to the other side of the shore. Sitting on the edge of Lake Melville, with the Mealy Mountain range in the distance, your eye is almost inadvertently drawn over the collection of houses that make up Sheshatshiu, one of Labrador's two large Innu communities. Prior to the construction of the bridge a small cable car ferried people and goods across the river in the winter months. Today, the two communities live shore to shore, with North West River ostensibly marked by the economic influence of Goose Bay and Sheshatshiu still struggling to claim (and redistribute) a portion of Nunatsiavut's emergent resource economy.[17] While the mission's materially fading legacy is easy enough to discern here, its efforts at sustaining, much like in St. Anthony, a sense of mutual cooperation, local industry, and an antiquated form of labor as self-help are distant indeed. When settler sites such as these came into being it was for their proximity to extensive trap lines to propagate the fur trade and for the shelter Lake Melville provided from the rough waters of the

Historic cable car line connecting Sheshatshiu and North West River, North West River, Newfoundland and Labrador, 2011. Photograph by author.

coast. It is perhaps an apt fate to have the flagship public museum of Labrador nestled away at the very end of the road in North West River. This is the northernmost point of a long tributary of the Trans-Canada Highway that branches off near Quebec City. At its one-way end, one finds the Labrador Interpretation Centre. As its name promises, interpretation abounds here; timelines have been made, Indigenous artifacts and traditions placed on low, stage-like pedestals, with the whole resembling an earnest, if impossible, attempt to recapture the sense of time going out of a distant world.

Down by the shores of the North West River stands the trapper's monument. Each face of its base is a testament to the families that lived and worked in the town. South face: "Many trappers had similar names"; East face: "Hudson's Bay Company Post Managers"; West face: "International Grenfell Association Staff"; North face: "North West River Trapper Monument." A metallic trapper checking his line strides above them. This too is a place of first fish—the coast that then called trappers into Central Labrador's interior. The trapper walks parallel to the river, being led toward the promise of the line.

North West River Trapper Monument. North West River, Newfoundland and Labrador, 2011. Photograph by author.

I met Dorothy McNeil at the St. Anthony Public Library in November 2011. I was looking over its small archival collection of mission-related documents, and I overheard a woman and a boy speaking French. The woman was patiently running through basic grammar exercises, overseeing what was clearly some post–school day tutoring. At the end of the lesson the woman came up to me and asked, "What are you doing?" From there, Dorothy would relate the story of her father, Edgar McNeil, and, after learning that I was interested in communications and the mission, the ways in which she understood how connected to the outside world St. Anthony was and is. We agreed to meet at her house, in the shadow of the Charles S. Curtis Memorial hospital, on the afternoon of the following day. Just before she left, Dorothy explained how she was helping the little boy catch up with his French class. He and his parents, both doctors at Curtis Memorial, had recently arrived in St. Anthony from Iraq.

The following is part of Dorothy's story of the ways in which living in St. Anthony was a networked condition or, as Dorothy puts it, how the residents of northern Newfoundland and Labrador "always sort of knew where was what in

Grenfell Mission hospital, St. Anthony, 1959. Photograph by Larry Davis, from "Historic St. Anthony Photos," Town of St. Anthony website, accessed March 12, 2014, http://www.town.stanthony.nf.ca/Historic_Photo_Gallery.php.

the world." It also offers a glimpse into Edgar "Ted" McNeil's life and times, not to mention the afterlives of his work as the mission foreman.

While sitting in Dorothy's living room late the following afternoon, I had a view out toward Curtis Memorial. Her constant stream of words flowed over me, and I had the privilege of just quietly listening to her evocative memory work. The light was gradually moving out of the sky, and the hospital loomed over us from its shadows. When Dorothy mentioned that the hospital her father had had a crucial hand in helping build in the 1920s was still standing, and then gestured at a nondescript building clad in blue sheet metal now used as the laundry for the main hospital, it was as if the sedimentary ties of infrastructural mediation fell into place. The long echoes of settler infrastructure building seemed to emanate out from the original hospital's metallic shell. Hospital built on hospital, a slow accretion of experiences that began to bind this settled place together. It was unsettling and suggestive of the living voices that continue to bring feelings of belonging, connection, and proximity to the way it was.

We weren't isolated.

First of all, my father, in 1928. I'm pretty sure Marconi sent someone out here, and he was trained. The mission was given a call number, and my father was trained to do the radio operating. . . . He would be the one communicating with the Boston office and New York. I remember when he first started he'd just send out a "CQ"—I suppose that's what you'd call it—that's when you call out and anybody can answer you, and eventually some man in Boston said, "Look, we'll have a regular schedule, really early every morning," and any messages would be passed along. If you had staff here from the States, they'd go over and give daddy a message or something, and he would send it off by talking to this man in Boston by voice or Morse code. . . . Daddy did it for the mission up until the war.

———————

To get down to mom's home [by boat], in St. Lunaire, for me in the summer that would be really thrilling. I grew up here, and it's kinda a different place; just walk down the road to this Grenfell school and back; the hospital was sort of right there, the focus. In those days it was only that blue one [points out the window]; it didn't have that facade then. It was an architect from New York who donated the design, and my father put it up. It was a really big do. This was 1925. That's when he went out to New York for the winter. He went over to the firm of Delano and Aldrich. The Rockefellers' house on the Hudson was built by them. Anyway, he went over there every day for about three months to learn how to construct the hospital 'cause it was steel beam. That had never been done, especially in northern Newfoundland. One fellow from here told me that the way the beams were put together was very advanced, even for Canada at the time. I don't know. There are two ways to do steel beams. That's all I know, really. I don't know what the two ways are. I got pictures of this. They dug the base, the foundations by hand. By horse and buggy. I think my father rigged up things like elevators and stuff like that for lifting things once they started on the second floor. Then they eventually had a humungous boiler come, and he figured out, with a pulley system, how to tug it up. But in the pictures it's a stone-colored brick façade, and some man did all that by hand. And then it was done to the specifications of Delano and Aldrich. Somewhere around the windows there was this pattern. So that's all covered in now. . . . The back part was built in the early 1950s. It was a TB san [sanatorium]. . . . My father drew up the plans for that. That part that's attached [to the original hospital].

I never knew anything about this, but my half brother told me, in more recent years, that when daddy was there in New York they wanted him, once he finished that project, to come back and work for them. 'Cause he picked up things very fast. A very quick learner. And he refused. He was very much a perfectionist . . . in building. Some man I met on a ferry once, who came from somewhere else in Newfoundland, before the war, this was the late '30s, he said he'd been at school with my father, and he said he could look up at a ceiling and see if it was half an inch off.

So anyway, daddy drew the plans for the san. Basic drawing he could do and carpentry. That furniture there, the table, and those over there [points to side tables], he did. The bookcases. That thing the TV is on. So, anyway . . .

There's too much to talk about . . .

I just found a postcard he was writing to mother. Or a little letter he wrote to mother, when he was in Ottawa. . . .

I remember one Christmas. We had an office in the house; it wasn't this house. I can remember him at the drawing board table. Always at the drawing board. So he got his plans approved when he was in Ottawa. Dr. Curtis was there, fussing and worried he wouldn't get funding for it. Or his plans weren't good enough. Or whatever. Anyway, they were fine. And so they got funding for the TB san . . .

How did I get off on that? . . .

Back to travel.

The coastal boats. . . .

———————————

People assume how isolated we were. But, as I told you yesterday, daddy was born in 1884, between Makkovik and Hopedale in a place called Island Harbour. It's the family homestead. When he was a child in the summertime they went out to some islands off shore, the Turnavik Islands, and so they could send the fish directly to the Bartletts because they were the ones who had the fishing premises. And so Grenfell came there in 1892. He stopped at Turnavik. Daddy would have only been about eight or something. I think he was there when Grenfell first came by. But Bob Bartlett . . . you really should read the famous story of his walking across to Siberia, but I won't get into that . . . there's so many stories. Bob was about sixteen to daddy's eight. So he was around. And I told you daddy would remember loading fish on the boats going to Spain, and I don't know if they were Newfoundland companies. I think they were. I don't know if the Bartletts had their own boat, schooners that took salt cod directly to Spain or Italy, I'm not sure; it could have been both. So here you are with boats going back and forth to Europe, and then the Moravians were there. So you were exposed to the world through them.

My great grandfather was living in St. John's. Some summers he'd come down from St. John's, give them news, and bring things for them that they couldn't get ordinarily. But there was always the Moravian, the German connection. Somewhere there was news getting around about what was happening in the world.

[. . .]

We had a visitor one day when I was about seventeen. He was daddy's age. He was a great friend of Grenfell's. He was Grenfell's best man at Grenfell's wedding in 1901 in Chicago. And he married Woodrow Wilson's daughter in 1912, and Grenfell was his best man at the White House. My half brother again told me, this was so long ago, that daddy apparently had an invitation to that wedding, but I don't . . . I never knew about it. . . . I didn't pay attention; like I say, when I was a kid, I wasn't one bit interested in all this . . . romanticism to me was living in France or England or Europe somewhere . . . it certainly wasn't this. I was like an only child. 'Cause my brother was so much older, and he was never home. So anyway . . . I just immersed myself in escapism . . . books. On the radio we'd hear good stories.

In one sense they weren't mentally isolated. And they always sort of knew where was what in the world. It hadn't been long before, when daddy was a little boy, his great grandmother was still alive, her husband was the one who came from Scotland. . . . So he had people coming, immediately, not that long ago from Scotland. It was not like a hundred years ago. And then on daddy's mother's side, his grandfather came from Norway. And he lived in Makkovik. He was alive until daddy was twenty. Daddy would have known somebody in his family who was from Norway. So you had that outside world connection. So you knew your basic geography, right.

Those Moravian missionaries had often been in South Africa. Daddy's teacher, I think that's where he met his wife. I think he was in South Africa as a child.

So, then . . .

Here's when I was seventeen in our house. Daddy was an old man, retired, and not very well. Here was this visitor in the house. I'm going back somewhere else now, back to the White House connection. A president's son-in-law coming in to visit . . . the same guy had set up the constitution, or helped do it, or drew it up for . . . the constitution or body of laws for Thailand. I got his book here somewhere. He ended up being a lawyer. He was a law professor at Harvard. I don't know if it was the prince or king of Thailand, it must have been his father; he went to the States and lived with this family. Anyway, after that this man was also high commissioner to the Philippines when war broke out, and that was McArthur, the general, they had to escape in a submarine from this place down the bay from Manila called Corregidor . . . that's a question in Trivial Pursuit, and I always know it. . . . Yes, the original Trivial Pursuit, and I always know it! That sticks in my head, the name "Corregidor." Now Wilson's daughter died, then he remarried, this man. He helped draw up things for the United Nations. His name is Francis Sayre.

———————————

I was going to show you this.

[. . .]

1. THE PLANT

On the ground floor of the Grenfell Interpretation Centre, in a three-foot-by-three-foot glass-topped case, sits a scale model of St. Anthony circa 1930. The town's topography is expertly rendered, with the mission's various structures, save for the materially distinct reinforced concrete hospital built in 1927, re-created with their white wooden facades and their windows with green trim. As you walk around the model, details emerge that strike you by their accuracy: the absence of sidewalks on the rough gravel roads, the basic telephone poles that are sparsely scattered, and the neat lawns bordered by thick-planked fences. The whole is so serene and pastoral it could be a model Swiss village on the North Atlantic.

Above the case is a black-and-white aerial photograph of the scene that presumably served as a tool in the making of the model. Each structure in the photograph is itemized and listed on the right-hand side. Of the thirty structures that appear in the photograph, twenty-five belong to the mission. Ranging from the mission store to the machine shop to the oddly paired power plant and library, this material infrastructure of the mission was a real-world testament to the practical philanthropic work the mission was there to undertake.[1]

Even beginning to describe the material extent of the Grenfell Mission is a difficult task. As a missionary enterprise with several decidedly reformative aims, the most important being shifting the fisherfolk's economic regime toward cooperative action, which is addressed in chapter 2, it took the matter of building seriously. What makes this description of the mission's material infrastructure difficult is that it was constantly changing and adapting to its circumstances, becoming an evolving, vital organism that morphed through a

FIGURE 1.1. Scale model of St. Anthony circa 1930, Grenfell Interpretation Centre, St. Anthony, Newfoundland and Labrador, Canada, 2011. Photograph by author.

complex interplay of disaster (especially fires), need, and the infusion of new fundraising capital. Grenfell's nearly annual fundraising tours throughout Canada, the United States, and Britain were meant to bolster the various national and regional Grenfell associations' base of supporters that ultimately would contribute, through small donations to substantial endowments, the vast majority of the mission's operating budget. While part of that budget went toward the purchase of medical and everyday supplies for the mission's hospitals, staff, and volunteers, much of it went into the boats and constantly growing collection of buildings that were needed to achieve its medical and social aims.

One of the most common ways that the mission's most influential actors, namely, its doctors, referred to the material embodiment of the mission was as a *plant*. This now slightly antiquated usage of the word was apt both for their intent as medical missionaries and in their purposes as colonialist frontiersmen and women bringing the benefits of organized society to the fisherfolk and Indigenous residents of northern Newfoundland and Labrador. The three most pertinent definitions of the term as a verb make up and allude to a complex of actors that reflect the constitution of the Grenfell Mission as a materially reforming colonial enterprise: (1) "To found or establish (a community or society, esp. a colony or church)"; (2) "To settle (a person) in a place; to establish (a person) as a settler or colonist"; (3) "To form a colony or colonies; to colonize, settle."[2] However, the most common way in which Grenfell Mission doctors, such as Harry Paddon, deployed the term was as a noun and as a sign of the mission's physical extent: "Our plant was primitive in the extreme. A square building, almost flat-roofed, about thirty feet by twenty-six, had to do duty for staff and patients."[3] Or, as Grenfell puts it in the IGA's first annual report in 1914, "the plant everywhere has been maintained," going on to update his benefactors on the progress made on the new store and wharf in St. Anthony as well as improvements to the water supply.[4]

Among its various early usages, the most illuminating occur in its varied life in the Bible: from locales destined for living,[5] populations available for divine settlement,[6] human-plant metaphors,[7] and fundamental creation stories[8] to divine intercession in human geography.[9] As both noun and verb, *plant*, though in common usage at this time, also foregrounds what can occur when religious belief, missionary practice (especially of a "preventive and curative" kind), and a complicated settler colonial situation combine to create the conditions under which planting as building would seem to offer the best possible solutions to make the best of possible (colonial) worlds. In what follows I seek to understand how the Grenfell Mission built that plant. I argue that the plant is an apt concept to address the mission's establishment of a form of settler infrastructure through the design, building, and maintenance of its early network of coastal medical care. This reform work undertaken by the mission was a form of infrastructural mediation predicated on the relationships mission employees established among hospital buildings and ships, the inhabitation patterns on the coastlines of Newfoundland and farther north in Labrador, an open-ended process of aging and constant upkeep, and an investment in a future-oriented religious affect defined by the Protestant homiletic tradition's emphasis on representable good works.

The infrastructural story this chapter unspools echoes Steven Jackson's framing of an "ethics of care" that lies in the repair work particular peoples at particular moments in history have given to new media technologies. Jackson's emphasis on a radically contextual understanding of *repair* is one that privileges a mode of analysis that looks beyond those dominant poles of sender and receiver, thus refocusing attention on the timely practices through which processes of mediation become sustainable and functioning and that can give primacy to the emergent relationality immediate users establish with their contemporaneous technologies.[10] This approach makes of the colony of Newfoundland in the year 1892 an emblematic site of mediated becoming, and it highlights how the mission projected an infrastructural zone extending out from the colony's social conditions. The ethics of care articulated by the mission's settler infrastructure was built on an evangelical Protestant tradition that privileged moral and purposeful action: here, infrastructural work entailed the constant projection of a material standard that would bring the fisherfolk in line with the care given to the mission's plant. Stemming from a religious lexical context, the plant's constituent elements also came from the early advances of international finance, the mission's diverse networks of (largely) Protestant philanthropy, new building standards and models imported from the northeastern United States and Britain, and the ways in which the mission's fundraising strategies combined them all to get such a diverse body of institutions planted in place on the remote coasts of northern Newfoundland and Labrador.

While rudimentary and often ill- and mistakenly equipped to perform in the subarctic environment, the Grenfell Mission's early buildings were a testament to its capacity to enact processes of living mediation in order to assure the success of their medical, social, and settler enterprise. If the mission was indeed the principal institutional actor making northern Newfoundland and Labrador into a viable site of resource extraction for late nineteenth-century settler colonists, then it follows that it had to establish a settler infrastructure of philanthropic concern from which to undertake its reformative goals. Opening up the physical plant of the mission to individual donors, as I address in the final section of this chapter, made the act of mere building into an ongoing and collective action of "planting," which, in turn, given its grounding in microscale Christian moral action of this world, required ongoing infrastructural upkeep, maintenance, and repair—a tending of both morals and materials that could be taken up into the Protestant homiletic tradition by the mission as a whole.[11]

"New Conditions Must Be Met in Every New Place"

It was in *Among the Deep Sea Fishers*, also referred to as the mission's journal, that donors were informed of the mission's various activities. Every issue was a combination of firsthand volunteer reportage; fundraising calls by different mission doctors, nurses, and administrators; uplifting narratives of need vanquished or disaster averted; natural historical treatments of the mission's geographic contexts; official reports; and a range of documents, such as photographs, maps, and poems, that could contribute to giving its readership an idea of the mission and its urgent missionary work. The publication was also the principal media interface for keeping track of the mission's built state. In nearly every issue, a recurring section titled "The Mission Stations" reprinted letters from doctors in the field at the mission's stations across northern Newfoundland and Labrador. These letters portrayed the day-to-day operations of the stations, occasionally punctuated by an acute or difficult medical case, but also offered a kind of running commentary on the physical state of the mission in its far-flung outposts on the Northern Peninsula as well as the South and North Coasts of Labrador.

At the scale of the mission as a whole, the "plant" referred to an impressive collection of medical, social, cultural, economic, and transportation infrastructures. This plant shifted over the course of the mission's evolution, from its lone medical ship of 1892 to its steadily growing collection of hospitals, nursing stations, and industrial centers. The term was nonetheless a pragmatic way of grounding the material infrastructure that the mission had fundraised for, bought, and ultimately built and maintained. In an October 1923 issue of *Among the Deep Sea Fishers*, the newly named executive officer of the IGA, Colonel Arthur Cosby, giving his "Impressions of the Mission Stations," reiterated the common geographical refrain shared by so many of the mission's workers and volunteers and a definite part of its broader North American and British cultural mystique:

> I was greatly impressed with the extent of our primary function—the Mission's medical work—and the manner in which the demand was being met. Look at the map at the back of this magazine. Note the long, tortuous coast line stretching hundreds of miles. Then remember that for the greater part of the year all this territory is ice-bound, absolutely cut-off, isolated. Remember, too, that in the few remaining months transportation is by boat only, often precarious with heavy seas, fogs, "dirty" weather, rocky shores. Remember also the small, thinly scattered population. Not a city, not a road, exists in all that country. Then only can you begin to visualize the Mission problem along the line of medical care alone.[12]

This visualization of the "Mission problem" was partly what different pieces in *Among the Deep Sea Fishers* were after: to provide a mental map of the mission (as a supplement to its various cartographic incarnations) that could bridge its supposed remoteness and empathetic distance and bring its distant and always tortuous coastlines a little closer to home and easy recognition. The map at the back of the October 1923 issue is an exemplary case of the mission's visualization strategies. The coast of Labrador takes up two-thirds of the map, imposing itself over the Northern Peninsula of the island of Newfoundland. The title of the map echoes this spatial distribution, "Labrador and Newfoundland," while also going against a common trope in the imperial imaginary, that of the invisibility of Labrador. Here it seems like it is a philanthropic cartography at work, with an ice-bound, isolated, and dangerous Labrador making its presence felt and, to some degree, justifying the denotative graphic scheme that shows Grenfell Mission stations marked by a star, Moravian mission stations with a black dot, and cooperative stores with overlapping black dots.

Take a close look at the map; it does not make for legible reading. While the words entirely in upper-case letters are mission institutions, they seem to meld with the myriad place-names. This pictorial system of reading, bound up in cartographic specificities, makes for a distinct impression of the immersion of the mission in its regional spatial context. You can follow the small stars up the coast of Labrador, marking the migratory fishermen's routes as well as, at their northernmost point, the unmarked line where settler populations give way to Inuit communities and the Moravian sphere of influence. A useful contrast in modes of philanthropic visualization and representation is to compare the 1923 map with the IGA's 1914 annual report. The visual "accounting" being done here is that of a quantifiable list showing number of patient visits, expenditures, and other monetized costs of operating the plant.

Cosby's narrative recounts the "tour of inspection" he undertook with Grenfell over the course of the previous summer.[13] They began at Harrington, in Canadian Labrador, looking out over the Gulf of St. Lawrence, and made their way to the town of North West River in the Labrador interior and then back down to the various stations on the Northern Peninsula. At each station he generally gives a first person account of the situation of the mission hospital, its staff, its surrounding region, and its place in the community. The descriptions display a domesticating tendency, picking up on the human-made objects that have beautified and psychologized the landscape: "Our next post is Forteau. The Mission house, freshly painted, surrounded by an attractive well-fenced garden, is fascinating. In many ways it is the most likeable house I saw on the whole of the Labrador coast. The bay is picturesque—indeed, rather romantic just now with the wreck of the British battle cruiser

FIGURE 1.2. (*facing page*) "Map of Labrador and Newfoundland Prepared for Dr. Wilfred T. Grenfell, C.M.G.," *Among the Deep Sea Fishers* 21, no. 3 (October 1923): 109.

65°   60°   Longitude West from Greenwich   55°   50°

LABRADOR
AND
NEWFOUNDLAND
PREPARED FOR
DR. WILFRED T. GRENFELL, C.M.G.

Scale of Miles

0   25   50   75   100   125   150   175   200

★ INDICATES DR. WILFRED T. GRENFELL MISSION STATION
• INDICATES MORAVIAN MISSION STATION
■ INDICATES CO-OPERATIVE STORE

UNGAVA
PORT BURWELL
KILLINIK
BAY
Cape Chidley
Grenfell Tickle
Johnut Inlet
Cape Kakkiviak
Eclipse Harbour
False Bay
Aulatsavik Island
Cape Macgregor
Cape White Handkerchief
NACHVAK Nachvak Bay
Cape Naksarektok
RAMAH
Hebron Island
Nisbet Bay
Big Island
Cape Ulvuk
HEBRON
Watchman Island
Cape Mugford
Nanuktuk Island
OKKAK
Saddle Island
North R.
PORT MANVERS
NAIN PAULS ISLAND, MISSION CHURCH,
Anatwaban R.
ZOAR
Ukasiksalik Island
Cape Harrigan
Nunaksaluk Island
Smallstone
Turnavik
HOPEDALE ILLIK
Cape Mokkovik
Kiallatuit Island
ADLAVIK
Webeck Harbour
Cape Harrison
False Cape
Byron Bay
Cut Throat Sound
INDIAN HARBOUR HOSPITAL
Hamilton Inlet
WEST BAY
Cape Porcupine
Gready Island
Spotted Island
SCHOOL AND DISPENSARY
Fishing Run
Seal Island
Boulters Rock
St. Michael Bay
Alexis R.
St. Lewis
Cape St. Lewis
BATTLE HARBOUR HOSPITAL
Great Caribou Sound
Cape St. Charles
Wireless Tel. Sta.
Belle Isle
INDUSTRIAL STA.
CO-OPERATIVE STORE
Windsor Tel. Sta.
GREAT CO-OPERATIVE STORE
ST. CAROL'S SCHOOL
ST. ANTHONY HOSPITAL, CHILDREN'S HOME,
IND. SCHOOL, CO-OPERATIVE STORE
St. Julians Harbour
Cape Harbour
Croix Island
Bell Island
Canada Bay

North R.
NORTH ATLANTIC OCEAN

Resolution Lake
George River
Michikamau Lake
Seal Lake
Fremont Lake
Mackenzie Lake
Windbound Lake
Lobstick Lake
Hamilton River
Minipi R.
Gose Bay
Lake Melville
Sandwich Bay
Grand Lake
Attikonak Lake
Burnt Lakes
Renonmtu R.

Q U E B E C
PAULS RIVER NURSING STATION
FORTEAU MISSION STATION
PAULS RIVER NURSING STATION
Twin Is. BODDICKTON
St. Johns SAW MILL,
Pt. Rich. STORE,
Keppel Island AGRICULTURAL
STATION
Hawke Bay
Portland Cove
Cape Whittle HARRINGTON HOSPITAL
St. Barbe or Horse Island
Partridge Pt.

MINGAN

ANTICOSTI

GULF OF
ST. LAWRENCE

St. Paul's Bay
Green Pt.
Bonne Bay
WINTER POST OFFICE FOR ST. ANTHONY
MAIL BY DOG TEAMS.
Bay of Islands
Port au Port Bay
Cape St. George
St. George Bay
Cape Anguille
Cape Ray
PORT AUX BASQUES
Cabot Sound
Cape North

Magdalen Islands

PRINCE EDWARD
ISLAND
CHARLOTTETOWN

CAPE BRETON
ISLAND

PILLEY ISLAND
HOSPITAL
LEWISPORT
NORRIS ARM NOTRE DAME JC.
Sir Charles Hamilton Sound
Green Bay
Notre Dame Bay
MILLERTOWN JUNC.
MILLERTOWN
Red Indian
Lake
Mary Harbour
Brookud Lake
SHOAL HARBOUR
Jubilee
Lake
Hermitage Bay
Miquelon
Islands
St. Pierre Island
Cape St. Mary's
St. Mary's Bay
Trepassey Bay
C. Pine
C. Race

NEWFOUNDLAND

Bonavista
Bay C. Bonavista
BONAVISTA
Trinity Bay Grates Pt.
C. Francis
CARBONEAR
PLACENTIA
Placentia
Bay
ST. JOHN'S
SEAMEN'S
INSTITUTE

C.S. HAMMOND & CO., N.Y.

60°   55°

RALEIGH just opposite."[14] Cosby maintains this domesticating, bucolic lens, describing the "cheery red roof peeking out from the heavy green foliage" at Lewis Bay, with a "tidy path, bordered by stones," the whole giving the effect of an "orderly compound."[15]

Cosby's tour was a sort of *état des lieux* of the mission as a whole. Checking up on places and people, the physical and mental health of buildings and minds, the tour also highlighted in its recurrence the need for continued donations. On a very mundane level, the plant, in its evolutionary state, could either grow or decay, and it was up to mission administrators to maintain the structural conditions of fundraising and the organization of labor and materials needed to ensure its ongoing vitality. "The upkeep of all the stations," as Cosby writes, "is a very considerable but most necessary charge on the Mission."[16]

Of the innumerable actors who made up the Grenfell Mission, none were perhaps quite as indispensable as those charged with the work of maintaining its physical infrastructure; they could be called *plant workers* for lack of a better term. Edgar "Ted" McNeil, Wilfred Mesher, Sam Acreman, and Jack Watts were names associated with building on the Northern Peninsula of Newfoundland and the coasts of Labrador. Of the three, Ted McNeil was probably the best-known tradesman, and he played a central and symbolic role in building the 1927 hospital in St. Anthony.[17] Like other young adults of the region, McNeil had been sent to the Pratt Institute in New York to be trained in the building sciences, and it was a point of pride that the 1927 hospital could be built using only local labor. After a fire in the early 1920s destroyed the cottage hospital at North West River, Dr. Charles Curtis would express his satisfaction with the new hospital by writing in *Among the Deep Sea Fishers*:

> We assembled all the materials here, everything. In the middle of June, Ted McNeil went down with a crowd of men and laid the foundations, then came back to get the lumber, etc. We had a cargo landed here and all was cut according to specifications, then the WOP [workers without pay] went again; got there about mid-July, when the real work was started, and the hospital was finished September 12, complete with two coats of paint. It is very warm and has a wonderful basement. I know you will be pleased to hear that after some doubt St. Anthony, and especially Ted McNeil, has proved that they can build. It is a record and everyone there is greatly pleased with it.[18]

Mission administrators and doctors, actors at the top of the mission hierarchy, did not take for granted that the plant had to be maintained. Provisioning the hospitals, cottage hospitals, nursing stations, and cooperative stores with supplies, both medical and such everyday essentials as food, clothing, and the like, was a constant preoccupation—a part of the mission's commitment to "infrastructuralism" and the logistics of care.[19] Yet it was Grenfell who thought most creatively of what provisioning the local population meant, scaling the practice up to encompass the mission's attempts to sustain the regions as a productive resource frontier and thus including initiatives such as cooperative finance and McNeil's time at Pratt. However, at the intersection of logistical work and the shoring up of future infrastructural arrangements, Grenfell, in his later life, hatched the scheme of bringing a stone craftsman to northern Newfoundland and Labrador:

> The houses of British subjects in the colony are deplorable and dangerous. I have slept and lodged in too many not to know it, and I have begun the use of concrete and teaching our people how to use it. We, however, felt that the endless stone all around us, and the limestone especially, would be used by the people if they knew how, and if only I could afford it I would have had out an expert long ago to teach us how to develop the use of that source of material, and to give courses in our schools in that and many technical arts and crafts.[20]

Grenfell's attempts at reform were decidedly practical, rooted in the very earth, or stone, that made up the ground of the mission's resident population.

By contrast, the mission's plant was predominantly made of wood, often transported by ship in ready-to-be-assembled pieces to the mission's various stations. As I will touch on in the section that follows, the mission's earliest seasonal hospital, that at Battle Harbour, was the result of a donation from a merchant firm and so not a purpose-built structure for the medical enterprise. It kept with the local wood-frame vernacular, giving even this most institutional of structures, a regional hospital, an unintentionally domestic, contextual, and anonymous aesthetic. As a result, and as this was a tradition of making do with donated infrastructure that would continue into the mission's future, retrofitting and the suitability of buildings for their varied purposes also were ongoing concerns for many of the mission's plant workers. Making materials both "fit" and "fit together" was a crucial task in creating the first nodes in the mission's establishment of a fully functioning medical infrastructure; "difference," according to Jackson, is revealed through the necessity of repair, and the mission's plant workers had to integrate this differential transition from colonial to missionary infrastructure.[21] Rather than an entirely evident case of "infrastructural inversion," to follow Geoffrey Bowker, the mission's practices of infrastructural mediation are suggestive of a middle ground of philanthropic practicality that held infrastructures in between the "figure-ground reversal" in order to maintain their suitability for repurposing and holding a symbolic as well as genuine value to the fisherfolk.[22]

A final challenge in characterizing the mission's plant is the absence of detailed descriptions of the infrastructures themselves. While many of the mission's boats have received careful historical treatment,[23] as have its two flagship buildings, the aforementioned 1927 hospital and the King George V Seamen's Institute in St. John's, many of the mission's nursing stations and cottage hospitals have been neglected by the archival record, and their presence in *Among the Deep Sea Fishers* is traceable only in their varied states of repair rather than in the particulars of their construction. In the following section I will attempt to piece together a more detailed narrative for the Battle Harbour hospital of 1893 as well as for the Indian Harbour hospital, which opened in 1894, but it is worthwhile to note here how, in the 1950s, insurance reports provided the most detailed if dryly accurate descriptions of the mission's plant. Giving the full details of construction, these reports provide another dimension to the infrastructural emergence of the plant: "*Machine shop.* Approximately 75 by 30 ft. with 30 by 30 ft. wing. Half-base made of concrete with concrete slab over same. Basement contains hot air furnace. Brick chimneys. Two-storey wood frame building. Walls and roof mineralized asphalt shingles. Located about 50 feet from hydrant."[24]

The Grenfell Mission plant was indeed an evolving condition that had to be constantly assessed in a timely manner, whether through material repair work on the infrastructures themselves or by shoring up the insurable dimensions of their medical missionary enterprise. "We are always growing," as Cosby is careful to remark in his narrative, "and new conditions must be met in every new place."[25] This maxim is indicative of the mission's commitment to an environment-responsive form of mediation. Improving the lives of fisherfolk entailed projecting a settler infrastructural work into the future. This recurring cycle of infrastructural care would, over the mission's nearly century-long existence, lead Dr. Gordon Thomas, the final iconic physician in the Grenfell Mission story, to state at the close of the mission in 1981:

> So it is with pride and good wishes that we turn over to the new Grenfell Regional Health Services Board the management and operation of a complex health organization and the transfer of our physical assets

worth in the vicinity of seventeen million dollars for the sum of one dollar. But it is not the physical plants that are important. It is the quality, the character, and the spirit of Grenfell which must continue to provide for the people of northern Newfoundland and Labrador the best possible service. Our prayers and Godspeed go with them.[26]

As befits a charitable enterprise, the mission's final donation of its physical assets to the province of Newfoundland and Labrador marked the integration of mission infrastructure with governmental jurisdiction.

Building under a "Fishocracy"

As Grenfell memorializes it in his autobiography, the genesis story behind the need for a hospital and trained nurse on the coast of Labrador came during his first trip "down North" on his arrival in Newfoundland in 1892. Following the fishing fleet of floaters and stationers that had already made the trip to the coast of Labrador for the summer fishery, Grenfell recalls arriving in the harbor of Domino Run. After seeing patients from the surrounding schooners for a full day, he noticed a decrepit boat off to the side, with a "half-clad, brown-haired, brown-faced figure" sitting in it in silence. Grenfell, observing him from the rail, was asked, "Be you a real doctor?," and was taken on shore to a "tiny sod-covered hovel, compared with which the Irish cabins were palaces. It had one window of odd fragments of glass. The floor was of pebbles from the beach; the earth walls were damp and chilly." With six "neglected" children huddling in a corner and a "very sick man" coughing harshly on a lower bunk while the "poor mother" tended to him with a spoon and bucket of cold water, Grenfell's portrait of outport poverty, while shaded with empathy-inducing descriptors, was accurate enough.[27] Ever the pragmatist, Grenfell's concern was not so much taken by the illness and poverty at hand; rather he saw that the man could not be taken away. "The thought of our attractive little hospital on board at once rose to my mind; but how could one sail away with this husband and father, probably never to bring him back." With the earning season already passing away, "while the fish were in," and the husband the only breadwinner, and so the entire family's welfare dependent on his labor, the hospital would have to come to him; a "hospital and trained nurse was the only chance for this bread-winner—and neither was available."[28]

It was this attention to distance and the necessary labor practices of the resident and itinerant fishermen that in part explains the spatial extent of the mission. Like many forms of seasonal labor, the Labrador summer fishery was an elastic enterprise of people, boats, harbors, trading stations, weather, local knowledge, and luck. It also traveled with the fish. When that enterprise went wrong—and even the smallest injury to a hand or arm could risk sidelining fishermen for a portion of the crucial season—its short duration became all the more apparent. Many of these injuries, especially minor cuts that could lead to infection, gangrene, and a debilitating amputation, were simply ignored and worked through in order not to lose a day of debt-relieving work. Hospital care, especially involving extended periods of treatment and convalescence, had to allow for proximity and speed. Grenfell treated some nine hundred patients off of, and on, the *Albert* during that first season of medical cruising. The ship was a mobile hospital that responded to the conditions imposed by the fishing industry; however, it was an equally itinerant service, which only made it all the more apparent that a fixed hospital needed to be brought to both the small population of *livyers* (year-round residents) and the floaters and stationers who worked off the coast of Labrador during the summer months.

With these conditions in mind and with the historical geography of the colonial fishery as a sort of infrastructural precedent, the island of Battle Harbour presented itself as the most favorable location for the mission's first hospital. Battle Harbour, which lies off the southeasternmost section of the Labrador coastline, became a crossroads and capital of Labrador of sorts over the nineteenth century by the fact that it was a literal safe harbor for the boats there to prosecute the Labrador fishery. A tourist today looking for that slightly eerie historical feeling of heritage accuracy, can spend a few nights in the restored Grenfell Doctors Cottage or former Royal Canadian Mounted Police detachment building, eat in the dining hall in the former general store, and just generally feel and touch the mercantile saltfish industry's material heritage that dates back to the eighteenth century.

The history of Battle Harbour reflects, at a reduced scale, the structure and actors of the Newfoundland fishery as a whole. With only three corporate owners over nearly 230 years of occupation, its final decline came in the 1990s with the virtual elimination of Newfoundland's inshore fishery. Its digital incarnation, battleharbour.com, adds another dimension to its mediated past. This latest iteration of the island as a tourist re-creation, managed by Battle Harbour Historic Trust, is also a seasonal form of labor creation, as it operates only during the summer months. And yet for all that, Battle Harbour is a remarkable case of corporate philanthropy and an indication of the interwoven relationships between merchants and missionaries. As I will outline below, just as the first hospital was the result of mercantile benevolence, Battle Harbour 2.0 came into being as a result of a wholesale donation of property rights to the mercantile properties by the third and final owners, Earle Freighting Services of Carbonear, Newfoundland. The logo of the Battle Harbour Historic Trust harks back to the days when Baine, Johnston and Company of St. John's was one of the largest firms in the colony. Its design was derived from a stencil found on the premises. The stencil was used to put the Baine, Johnston and Company mark on the then-plentiful barrels, boxes, and other containers used to ship their fish to other parts of Newfoundland, Spain, and Portugal.

The island of Battle Harbour exemplifies how the coasts of Newfoundland and Labrador laid bare their own infrastructural conditions through their North Atlantic ecologies. Seen through settler eyes, whether livyer, stationer, or floater, these wind-shorn rocky outcroppings were sites to process, pack, and ship catches of fish: potentially fleeting, if necessary, places between the colonial metropole of St. John's and the waters of the Labrador fishery.

After Grenfell's season of medical cruising during the summer of 1892 and on his return to St. John's, he presented a report on the medical work he had undertaken. This report was received by an amorphous group of government officials, nongovernmental mercantile interests, and quite a few interested parties that blurred the boundaries between public service and private enterprise. St. John's society in the 1890s was divided by strict class lines, caste beliefs that stretched back colonialist generations, and religious ties that determined the extent of one's personal and, for the more orthodox, commercial world.[29] For Grenfell, still working under the auspices of the London-based Royal National Mission to Deep Sea Fishermen, public support in St. John's came in the form of government and mercantile assistance, with the Newfoundland government as well as the Water Street fish merchants promising to build and equip two small hospitals and also sustain a grant for their upkeep. The city's commercial interests, notably Baine, Johnston, pledged to give Grenfell a house at Battle Harbour and another at Smoky Run, near Hamilton Inlet. The merchants also agreed to contribute fifty to one hundred dollars each and also to assist with the logistics of patient transportation.

This display of altruism was of course anything but. One of the tagalongs on Grenfell's cruise in 1892 was Superintendent of Fisheries Adolf Nielsen, who quickly observed that a hospital not only would benefit the fishermen but also would ensure a reliable labor force and give some much-needed human insurance to the mercantile firms, passing along his observations in a letter to Governor Terence O'Brien: "Besides all this loss, suffering, and miseries a sick fisherman has to bear himself, he often also causes a great loss to the planter or merchant who supply him, when deprived from prosecuting the fishery, perhaps in the best part of the season, especially in such cases which nature takes a long time to heal, but which by professional treatment and attention very likely would have been cured in a short time."[30] Nielsen's final remark adds another temporal dimension to the need for a fixed hospital. A speedy recovery was contingent on proper medicalized care that could reinsert that fisherman back into an economy of which he was an integral if minor part. Grenfell had thus secured the public support that mattered, and yet, as Ronald Rompkey notes, it also placed the initial development of his medical missionary enterprise in the hands of the mercantile firms that had created both the structural conditions for and the working conditions of the itinerant and permanent fishermen.[31] The merchant firms, a veritable "fishocracy,"[32] as one British observer remarked, were both the problem and the class-driven solution to the fisherfolk's poverty.

In a trend that continues to this day in so-called privileged (that is, sanctioned by international government policy) developing areas of the world, it was medical intervention that managed to make it through public and private entrenched interests, those in whose hands the wealth was held, in order to improve, first, the health of the fisherfolk and, second, in time, their extra-labor lives. Like the proverbial wedge, medicalized aid could pry open the colony's monopolistic fishery.[33] As Ashley Carse emphasizes, "infrastructural work" is a relationship-building phenomenon that is both "social and ecological."[34] In this respect, the mission's early medical and, by extension, settler infrastructure was deeply bound to the colony's political economic conditions and their articulation of the exogenous development of the North Atlantic fishery.[35] Medical care, in this context, functioned as a sort of overlay on the preexisting geographies of the resource frontier, a case of missionary infrastructural mediation evolving as a socially responsive material form that would facilitate and become concretized as the mission's early medical infrastructure.[36]

The following summer, in early May 1893, Grenfell was putting the final organizational touches on his return trip to the colony. On the previous Labrador trip Grenfell had been the only doctor onboard. This time two volunteer physicians, Eliot Curwen of Cambridge, England, and Alfred Bobardt of Melbourne, Australia, and two nurses, Sister Cecilia Williams and Sister Ada Carwardine of the London Hospital, were to accompany Grenfell and undertake the opening of the two hospitals at Battle Harbour and Indian Harbour. In addition to being donations by two different mercantile firms, the aforementioned Baine, Johnston for the former and Job Brothers and Company for the latter, both hospital sites were also what Grenfell calls "'bring-ups,'" that is, more or less stopping places for the schooners traveling to and from Labrador.[37] Eliot Curwen kept a diary of his time in the colony, and it offers several insights into how the hospitals came about, including sketched floor plans. He was to take charge of the hospital at Indian Harbour along with Sister Williams, while Dr. Bobardt and Sister Carwardine were to manage the opening of the hospital at Battle Harbour. Curwen also took a series of documentary photographs during his voyage in Labrador. While potentially intended for Grenfell's fundraising activities that were to follow in St. John's, and in the United Kingdom, the photographs show aspects of the fishery as well as the range of populations, from temporary English traders to Inuit women employed by the Moravian mission, who lived and worked in Labrador.[38]

Another change from the previous trip was the composition of Grenfell's crew. The mate, second mate, and steward were all evangelical Protestants, and their numbers also included a carpenter, who was to assist in the work of hospital construction, a sailmaker, and three general seamen.[39] Grenfell's missionary labor force was part of his more elaborate preparations to change the medical landscape of Labrador. Given that the Battle Harbour hospital, or in its incarnation at the time as a merchant firm house, was not purpose-built but rather the result of corporate donation, as I note above, there is a haphazardness and ideology of making do in these early efforts to establish a medical infrastructure for the fisherfolk of the colony.

Unlike the precise and calculated outfitting and provisioning of the mission's boats (even en route to Newfoundland from England Grenfell made an unexpected stop in Bristol to receive the donation of a small vessel, the *Clywd*, that he had reworked for service off the colony's coast and then shipped via steamer to be of use on their arrival),[40] the matter of the bricks and mortar (or wood) of the medical operation only now became a matter of infrastructural concern. Hence the inclusion of a carpenter as part of Grenfell's crew and, at this early stage in the mission's history, building materials that could be part of a political economic negotiation between merchants and missionaries in order to secure support for the hospital projects. The infrastructural work Grenfell was spearheading was reliant on making multiple aspects of the colonial enterprise fit together. The settler infrastructure the mission was designing was a site of collaboration between merchant and missionary as well as a relationship predicated on a mutual understanding of and commitment to the necessities of repair. While this characterization may be true of the Battle Harbour hospital, by way of contrast, the Indian Harbour hospital, as I will discuss below, was a purpose-built structure with its constituent pieces prefabricated in St. John's in preparation for Grenfell's arrival.

As Curwen's floor plans show—as does one of his photographs that includes the Mission to Deep Sea Fishermen (MDSF) hospital, the agent George Hall's storehouse, and Hall's house—the Battle Harbour hospital, like so many structures in the colony, was an iteration of the local vernacular.[41] It was a house, not a hospital. Or it was a hospital as a house. Looking between the MDSF hospital and Hall's storehouse in Curwen's photograph, the two dwellings are essentially identical, save for the white picket fence around Hall's residence and a ladder propped against the front of the MDSF hospital, an indication that repair work was ongoing.

On the ground floor of the hospital were a kitchen, dining room, consulting room, meeting room, bath, water closet (toilet), and coal and wood storage. The second floor housed two wards, one for women and one for men, the nurses' room, a steward's room, and a convalescent room. Each ward was roughly fourteen by fourteen feet, while the convalescent's room, presumably meant for fewer people, was fourteen by ten feet. Like many, even metropolitan, hospitals of the time, this was a domestication of institutional medical architecture. As architectural historian Shane O'Dea notes, "it was the sort of domestic that speaks 'home' rather than 'house' with diamond-pane windows and bargeboards—the vocabulary of Olde England."[42] It was what was given to Grenfell and the MDSF, and so it was modified and made good enough. Curwen's diary description of Battle Harbour echoes this open-endedness of the local infrastructure, both social and material: "Battle Harbour—the Capital of Labrador—is an odd looking little place but picturesquely situated on the rock; it consists of a church without a clergyman, a schoolhouse at present without a schoolmaster, an agent's house with Mr. Hall the agent, a large storehouse, the hospital not yet fully completed and several fishermen's houses and wharfs & stages for drying fish; these buildings are scattered and as far as I have seen are unconnected by roads."[43]

Curwen's Battle Harbour is a place of both past infrastructure and infrastructure to come; the "living mediation" that Berlant sees as being at stake across infrastructural landscapes is reduced here to the presence of "Mr. Hall the agent." Yet Battle Harbour was not a straightforward instance of infrastructural failure. Rather, its settler colonial infrastructure is suggestive of the transient and seasonal nature of the Labrador fishery, at base its political economic strictures, and also of this broader condition of a merchant infrastructural capacity that was able to be repurposed. Just as Berlant sees infrastructure as sustaining instances of "proximity" that build toward a "world-sustaining relation," Battle Harbour's absence-filled church, school, and hospital push up against the longevity of the merchant's presence on the island and point to the precarity of the social relations underpinning the merchant-fisherfolk dependency.[44] It is also a state of affairs that parallels the broader phenomenon of the relative isolation of the colony of Newfoundland, and points to the fact that the colony lacked a standard of comparison by which to measure its own social and institutional living conditions. As one of Grenfell's early biographers remarks in his preparatory notes: "Because they lived in a small world of their own up to the time of Confederation, neither the Government nor people of Newfoundland had been truly aware of the full scale of the deficiencies in their public services. Their inadequacy by any standard was recognized. The size of that inadequacy, when new criteria emerged after 1949 in a study of the services of even the poorest Canadian provinces, was found to be staggering."[45] While this is a debatable claim, it is noteworthy given that it resembles the logic, writ large at a societal level, of infrastructural work undertaken by Grenfell and the MDSF. It was a matter of making do, getting it up, carrying on—of making this earliest manifestation of the plant a founding moment of acceptance of what was given and by whom—this meeting point of merchant and missionary around the politics of infrastructural repair.

As such, the designs of the mission's charitableness were materially haphazard, lacking a standard of comparison and aimed at preparing the ground for intervention, wedging economic disparities into a domestic, both regional and home-based, medical infrastructure.[46] What makes it a remarkable instance of imperial philanthropy is that the material donation came from within the colony itself. Grenfell's biographer, trying to both establish the context within which his actions could have taken place and account for the spur that made the man of venture, retraces the colonial narrative fate of Newfoundland:

> The story is unique. There is no parallel for the policy of retarded colonization. It was that policy which forced dispersal of the population along the perimeter of the Island and multiplied the problems of government to a degree unknown elsewhere in the civilized world. And the problems of providing adequate communications and reasonably good services, complicated by the French Shore controversy, would have been difficult of any solution by a much wealthier government or a much richer country. In the case of Newfoundland, it was not until 1940 that a degree of fi-

nancial capacity was reached that would enable the first important steps to be taken to remedy the wrongs and abuses of an extraordinary history which had caused a great English statesman of the nineteenth century to refer to Newfoundland as "the sport of historic misfortune."[47]

This "policy of retarded colonization" is precisely what eventually lured Grenfell away from the MDSF and toward the establishment of the IGA: it was the "lure of the Labrador" as well.[48]

Having to work against an established system of conditions by donation meant that the mission was also working on the synchronicity and geography of the region's standard of comparison. Paul Edwards's "modern infrastructural ideal" of the late twentieth century was, on this colonial margin, a practice of projection that sought to improve fisherfolk lives by modifying the resource frontier they had settled through an evangelical faith in material standards.[49] As the IGA developed, it was constantly bringing in, importing, establishing, and assembling equally ideologies, practices, and materials—a "living mediation" of infrastructural work in the making. As a result, the mission became structured around its practices of infrastructural mediation in order to manage that original "promise of extraction" through an infrastructuralism turned toward, in the early days of the missionary enterprise in the 1890s, medical care above all else. The resource frontier they were working across was a living environment that they saw as open to refashioning, particularly through the deployment of infrastructures—here of medical care—that could reflect international standards of comparison.

One of my favorite files in the mission archive, "Equipment on Coast—Data 1931–1948," is a collection of pamphlets from the period for such items as a "5 KW Kohler Battery Charging electric plant," a "Repair Part List for 8 HP. Tillavator," a Wincharger wind-power generator for radios, and a "new model 'AG' Bovie electrosurgical unit" as well as elaborate lists of the existing equipment at the various mission stations, such as St. Anthony's rock crusher, "Type 'B' Blake Crusher #2 Size 10" × 7,"" manufactured in Milwaukee, Wisconsin.[50] It was "retarded colonization" that played a role in making the fisherfolk of northern Newfoundland and Labrador able to be brought up to speed (and to standard) by the mission.[51]

The infrastructural work underlying the building of the Battle Harbour hospital demonstrates how the mission articulated its understanding of the plant as a colonial enterprise predicated on the maintenance and repair of their charitable infrastructure. In the broadest sense for these missionaries, the plant largely functioned as shorthand for the means of organizing the

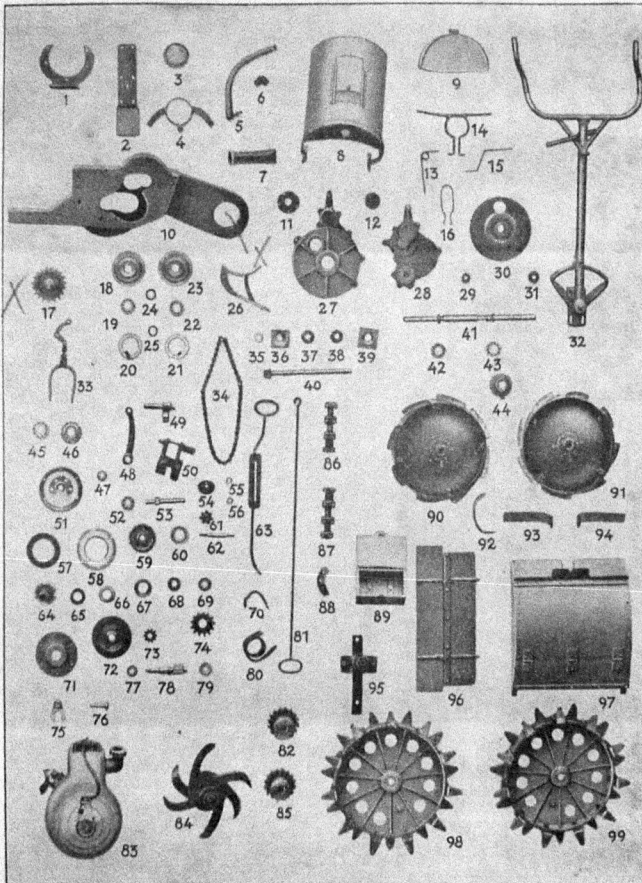

FIGURE 1.3. "Repair Part List for 8 HP. Tillavator," MG 63.1978, "Equipment on Coast—Data 1931–1948," International Grenfell Association Fonds, Grenfell Association of Great Britain and Ireland, Provincial Archives of Newfoundland and Labrador.

FIGURE 1.4. Wincharger pamphlet (n.d.), MG 63.1978, "Equipment on Coast—Data 1931–1948," International Grenfell Association Fonds, Grenfell Association of Great Britain and Ireland, Provincial Archives of Newfoundland and Labrador.

lives of fisherfolk, what Berlant terms the "lifeworld of structure."[52] This was the world the Grenfell Mission was trying to sustain in its earliest stages, thus their emphasis on infrastructural care and repair dependent on merchant capital. The story of the emergence of the mission's early medical infrastructure articulates how such historiographical work around the appearance of infrastructure-as-relation can lay bare the "patterns, habits, norms, and scenes of assemblage and use" that would come into being.[53] The consolidation of this settler infrastructure also foregrounds how media infrastructure studies can attend to the sedimented vitalities that lie submerged across past infrastructural networks, in this case, an evangelical Protestant investment in practical therapy that was made manifest as infrastructural work that included fisherfolk lives.

If the Battle Harbour hospital exemplified this infrastructural work of repair, that at Indian Harbour was part of the mission's effort to extend its medical network to the less capitalized shores of Labrador. The Indian Harbour hospital, as the second of the two mission projects, was a latecomer for Curwen and Sister Williams. While the plan had been for the doctor and nurse to move on to Indian Harbour after a few days at Battle Harbour, the mail boat had not managed to land the necessary building materials at Indian Harbour on its first pass. The boat had carried them up along to the North Coast of Labrador and only deposited them in late summer. While the foundations were laid on August 3, the second MDSF hospital was not completed until October and, so, well beyond the fishery season and on the cusp of a winter of immobility and inaccessibility due to the pack ice.[54] The photograph of the completed Indian Harbour hospital included in the contemporary published version of Curwen's diary was most likely taken by Grenfell at a later date. "We did not land at Indian Harb., as we wanted to get on & Dr. G. was going to call in the steam launch," Curwen writes; "we hear the hospital is finished and locked up, the carpenters having gone home. I should like to have seen it and seen some of the people again, but it is just as well we did not land."[55] While in St. John's, Curwen had, in his shorthand, met the "builder of my hospital, see[n] wood in preparation & talk[ed] over plans."[56] His rough plan of the ground and first floors measures the hospital's outside as extending thirty-two by twenty-two feet.

Although the Indian Harbour hospital was a rudimentary institution, it was also a part of the mission's medical infrastructure and thus subject to the local standard of measurement—a saltbox domestic vernacular. Built by local labor, it merely extended the original merchant-driven plant at the inherited building standard. Much like the arrangement of the Battle Harbour hospital, that at Indian Harbour incorporated two wards on the second floor, separated by the steward's room, a nurse's room, a common bathroom, and a dual-purpose dining room/convalescent room. On the ground floor, were a multipurpose waiting hall, a doctor's room and consulting room, the hospital's kitchen, and a rear addition containing the water closet (toilet) and stores.

On Curwen's journey home, after having seen 1,001 patients while cruising on the *Albert* over the summer months, he stopped off at Battle Harbour. "Our friends at Battle—Dr., nurses & Mr. Hall—we found well, and the hospital pleased me greatly; it has those characteristics of warmth, cleanliness and brightness that a hospital should have."[57] His work on the Labrador coast finished, Curwen could survey the homely hospital from a certain remove. Did the imperial and colonial governments' policies of actively dissuading permanent settlers and colonization over hundreds of years through selective land grants, violent expulsions of squatting populations, and restriction of fishing rights to floating populations catch the colony in a sort of time lag of material development? Or, did the seasonal ideology of the fishery dominate governmental action, or inaction, to such an extent that the colony as a whole could be viewed only as "'a great ship moored near the Grand Banks,' . . . for centuries officially regarded as no more than a training ground for the Royal Navy and an exclusive province of the West Country adventurers who took part in the summer fishery"?[58]

Between ship and shore, government and merchant firm, the Battle Harbour and Indian Harbour hospitals show how a missionary institution such as the Grenfell Mission could steer a state of "retarded colonization" toward its own infrastructural ends. John Durham Peters relates that the "ship is not only a metaphor; it is an arch-medium that reveals the ontological indiscernibility of medium and world."[59] For the late nineteenth-century fisherfolk of northern Newfoundland and Labrador, this "ontological indiscernibility" was at the foundation of the resource frontier they sustained and inhabited. This was the broad medium-environment that this early period in the Grenfell Mission's influence was inflecting through its infrastructural repair work. The resource frontier could serve as an environment capable of registering the material effects and social affects that came into tension through imperial political economic pressures and, in turn, the mission's attempts to create a livable regional economy. As I will touch on below, this ethos of cyclical infrastructuralism would guide the development of the material and evangelical Protestant practices of reform that the mission would undertake over the ensuing decades. In the 1890s the mission was operating across a "fishocracy" and guided by an inherited maxim of infrastructural repair work: "Home is where the medicine is."[60]

A Brick for Labrador

Writing in 1915, Dr. Harry Paddon, newly established at Indian Harbour, took up the hospital that Curwen was to have overseen in those early days of 1893. As with much of the mission's infrastructure, it had aged, and aged all the more quickly, given the harsh climate—with wind, snow, and poor if nonexistent insulation making aging an inevitable and recurring condition. The pride that Curwen had in establishing the designs for the mission's first purpose-built hospital had slowly degraded into Paddon's previously cited description of the Indian Harbour hospital: "Our plant was primitive in the extreme. . . . There was no plumbing at all inside."[61] Paddon's observations show that expectations, and those ever present if distant standards, had changed.

Paddon recalls receiving, two years prior to his arrival in Labrador, an "epistle" from Dr. Grenfell of a mere four lines, a bit of a forewarning of what to expect. "All knowledge is useful out here," Grenfell writes. "Learn all you can of house-building, plumbing, agriculture, stock-farming, etc., etc." Finishing up his medical degree at the time, Paddon was forewarned that his medical credentials were, as he himself observed, "only an item in the needful preparation for useful service in Labrador."[62] Those et ceteras would have only vaguely prepared Paddon for the scope of "doctoring" he was to encounter in Labrador. "How many times this summer," Paddon writes, "have I wished I were a certified engineer!"[63] There was always work to be done, and not all of it was of a strictly medical kind, though taken with a broad lens it was certainly infrastructurally and repair-driven with a view to supporting the ongoing work of the medical enterprise.

As Paddon's observation on the state of the plant makes clear, an all-encompassing outlook on the conditions of the mission stations generally prevailed. Remarking on the state of the plant rather than on the state of this building or that particular operating room reveals a panoptic concern for the physical infrastructure of the mission. As a settler colonial environment, the plant had to hold together across a disparate, remote geography and temporal markers of missionary presence, care, and influences. As such, newness and renewal of the plant as a unitary infrastructural entity were abiding concerns and, as I note above, accounted for many of the fundraising efforts undertaken by the mission. One of the preconditions of settler infrastructure in this context was its negotiation of such markers as aging and how they could be associated with the fading influence of an institution such as the Grenfell Mission. Even the idea of *newness*, especially as it pertained to the actual infrastructural work and things of the Mission stations, whether pharmaceuticals and clothing or water boilers and Ford trucks, was a means for the mission's administration, made most manifest in *Among the Deep Sea Fishers*, to maintain and garner sustained interest in their initiatives. They had to convince the donating public that there was a need for these new material objects, and none were so costly as the massive assemblage of materiality that was a fully outfitted outport hospital. The perceptions and realities of infrastructural progress had to be made to coincide. The challenge administrators faced was mapping their network of philanthropic concern over the plant as a whole—to, in many respects, make the infrastructural mediation the mission was there to undertake hold together across the resource frontier the fisherfolk worked for the benefit of the British Empire.

While many of Grenfell's religious writings, such as *What Can Jesus Christ Do with Me?* and *Religion in Everyday Life*, emphasize an earthly turn to action, they do so by holding up the material conditions of the world as a prime mediator through which such eternal values as selflessness, courage, and moral immortality can be made actual.[64] Resonant with the broader tenets of the then-prevalent social gospel movement, the mission sought to quite literally build religious ideals through material infrastructure. The mundane and earthly here and now mattered in Grenfell's system of missionary concern to such a degree that the mission became a nondenominational enterprise at an intentionally nebulous date soon after its incorporation. However, it was clearly an inclusive pan-Christian (if largely Protestant) espousal of material acts of charity. This Christian and settler infrastructure of care was rooted in its own open-ended doctrine of ministry. This outlook guided how the mission approached its infrastructural work of constant repair and renewal and, following Berlant, demonstrates that a particular social form (here, an extractive colonial capitalism) reliant on "resilience and repair" does not "necessarily neutralize the problem that generated the need for them."[65] Where Berlant sees the mobile patterning of infrastructure opening up social and political possibilities of alternative associations, repair work undertaken by the mission as a practical ministry constituted a genuine attempt to improve the working and living conditions of the fisherfolk, though only within the context of Britain's imperial fishery. The imperative of the plant as an infrastructural ordering of settler colonist-missionary relations relied on repair, in part, as a justification for the extension of the Grenfell Mission's presence into an undefined future. The very presence of these material infrastructures was a means of staking a homiletic claim on what kind of (Protestant, extractive, cooperative) settler life should be led. In order to sustain this ministry, the persistent question that Grenfell and mission administrators had to address was how to make bricks and mortar count—how to cyclically integrate infrastructural repair into the IGA's fundraising activities.

As noted at the outset of this chapter, it was in the pages of *Among the Deep Sea Fishers* that the mission's capital campaigns found their most comprehensive expression. A recurring fundraising trope is the brick as a charitable unit of measure. In a 1913 issue of *Among the Deep Sea Fishers*, as part of his Doctor Grenfell's Log series in which he recounts that season's toils and hardships—with patients' conditions related in detail—Grenfell describes the progress made on the projected Seamen's Institute in St. John's and asserts: "From the very bottom of our heart there goes out to those who paid for one brick or one pound of mortar the heartiest of hearty feelings of gratitude."[66] This parsing of progress was intended to make donations equitable and manageable among a varied group of donors. It was also a powerful image that replaced dollars with bricks and thereby turned that prime religious mediator into a concrete form of symbolism. While Doctor Grenfell's Log is immediately followed by a "Form of Bequest in Aid of the Work of the Labrador Medical Mission," which helped readers bookend their concerns by providing a clearly worded paragraph that could be inserted into their wills, the brick was a recurring trope in the life of the publication and an equally earnest call for aid that could be both annual as well as project-specific, as needs arose and the conditions of the plant dictated. There was not much calculation behind the appeal; editing was not undertaken with a view to manipulating their donors but rather to ask, as Grenfell bluntly does, "Who Will Provide for These Special Needs?" With effusive thanks, "that I should pose as Oliver Twist before Mr. Squeers," Grenfell asks for help to build a new wharf and a new cottage hospital in Indian Harbour.[67] The popular press, and most particularly the *New York Times*, given the mission's quick expansion into the northeastern United States, could and did mobilize public interest in the mission's work, especially when either tragedy struck or world economic conditions were dire: "While the burning of the hospital at Battle Harbour has done much damage, it has been as a light in the northern sky, reminding us of the work that is carried on in that inhospitable region."[68] Buildings, whether standing, to be built, or aflame, could serve multiple purposes.

One of the mission's publications that makes the case for the brick in an almost literal sense, is the booklet called "A Brick for Labrador." The cover gives the illusion of presenting a three-dimensional, cheery red brick. Published by

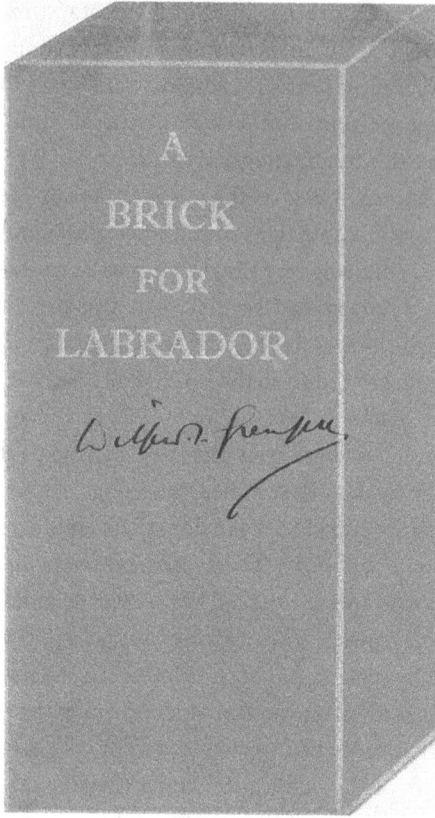

FIGURE 1.5. Lady Grenfell, "A Brick for Labrador" (n.d.), MG 63.2207, Grenfell Mission Leaflets and Booklets, New England Grenfell Association, International Grenfell Association Fonds, Grenfell Association of Great Britain and Ireland, Provincial Archives of Newfoundland and Labrador.

the IGA of New York, printed and assembled (not a particularly easy process) by the students of Hampton Normal and Agricultural Institute in Virginia, written by Lady Grenfell, and featuring a foreword and drawings by Grenfell himself, the booklet sold for twenty-five cents, with each one described as a "contribution to the work in Labrador."[69] The text making up the contents of the booklet—semi-biblically titled "The Land of the Naked Rock"—treads common ground, from the Vikings to the colony's vast natural resources to Labrador's potential in the bourgeoning tourist trade to the mission's genesis story for the prevailing conditions in the colony of Newfoundland. It gives a comprehensive overview of all mission activities to date, ranging from its educational programs for local children to learn technical trades to its industrial work in toy and clothing manufacturing and including a point form list of its hospitals, nursing stations, and industrial centers. Its main purpose is to serve as an informative fundraising tool—to make donors aware of the mission's work, its very large impact requiring increasing support, and the scale of the ongoing, seemingly open-ended needs of its ministry. The main body of the text ends with a reminder that, for three dollars a year, readers can become members of the Grenfell Association of America, which also comes with a subscription to *Among the Deep Sea Fishers*: "This periodical brings the 'mountain to Mohammed' by keeping Labrador's friends in close touch with the activities of the north. Your membership will go directly to the upkeep and support of Sir Wilfred's work. Won't you help the Labrador in this way?"[70]

The booklet as a whole is an effective charitable object, and it embodies the literal sense of infrastructural visualization that partly defined the mission's outlook on the material good works of infrastructure. A single brick was of course a metonymic unit that had to be made to echo the singular individual's spiritual donation and concern while also standing in for a collective agency of need (in this case, "Labrador"). While the pamphlet lacks a date of publication, it can be dated to sometime after 1927, as it alludes to the successful completion of a "splendid new hospital, built of concrete blocks, and equipped with every modern convenience for the safety of patients and the comfort of the staff,"[71] and so picks up and perpetuates the brick trope after it had, to some degree, come to its fullest fruition. Bricks could also serve as a reminder to mission donors that the very existence of fisherfolk depended on the material presence they were sustaining; as Peters notes, ontology "is usually just forgotten infrastructure."[72] It was up to the mission journal to foreground how the ontological and the infrastructural were central tenets of their settler colonial ministry. Settlement of the colony of Newfoundland was an open-ended project that had to prioritize settler lives over all others. The mission sought to build its own conception of critical infrastructure that derived from its ontological commitment to an evangelical Protestant worldview that could sustain the British Empire.[73]

It is difficult to track the mission's varied brick campaigns through *Among the Deep Sea Fishers*. References to them tend to come under such headings as "The Tale of Bricks," with a list following below that tallies up the names and number of bricks (at twenty-five cents each) donated by an individual. It is a charitable and architectonic trope meant to appeal equally to children and adults. In the October 1918 issue of *Among the Deep Sea Fishers*, the Children's Page section features a drawing of a group of children ascending a set of steps beneath a brick arch made up of twenty bricks with the inscription "This Way to the Brick Home for Little Children at St. Anthony" inscribed above it. The accompanying text encourages young readers and their parents to collect

Children's Page

THE CHILDREN'S HOME, ST. ANTHONY

THE Editor is confident that all the young readers of this magazine, as well as the older ones will be greatly interested in "Dr. Grenfell's Log," especially in the part descriptive of the fire that came so near destroying the old wooden orphanage. Although this house is very cold and uncomfortable in winter it is the only home of more than thirty fatherless and motherless boys and girls in a little village comprised of but a few more wooden buildings. While reading of the narrow escape of Ettie and Johnnie, and Nathan and others, think of the long, long journey the children would have had to take, in open boats mayhap, along a very sparsely settled coast before they could all have found a shelter. Remember, too, the request of those same boys and girls that the boys and girls in the good comfortable homes in the United States and Canada and England, buy and ask their friends to buy 25 cent bricks to help put up a nice, warm, safe, fireproof brick building. Twenty-five cents to some of our boy and girl readers means only giving up some little extra luxury, something special to eat or to play with, a big self-denial; but if we are all really anxious and willing to help these interesting children of St. Anthony, we can all find a way of doing it. Will we make Dr. Grenfell's heart, as well as the hearts of these boys and girls glad by assuring them that by next summer we will send the CLUETT loaded down with bricks for the new Home? If the boys and girls will add the number of bricks printed in the various columns and multiply the total by 25 cents, the cost of a single brick, they will see that we have now enough money to buy 16,669 bricks. Dr. Grenfell says he thinks he can put up a brick building for about $40,000. We have now over one-tenth of the required sum. In the next six months let us double our efforts so that in April we may present Dr. Grenfell with the full amount of money for the new building. Will you do it; Grenfell Leaguers and friends? "A long pull, a strong pull, and a pull altogether" and the Labrador children will have their Home before another cold winter.

THIS WAY TO THE BRICK HOME FOR LITTLE CHILDREN AT ST. ANTHONY

Each Brick in Place costs 25 cents.
Will you help by sending 20 Bricks?

MARK OFF EACH BRICK IN RED AS YOU COLLECT THE MONEY.

When the arch is completed keep the picture but send this slip with the money and your name and address to

The Grenfell Association of America, Inc.
156 Fifth Avenue, New York, N. Y.

FIGURE 1.6. "Children's Page: The Children's Home, St. Anthony," *Among the Deep Sea Fishers* 16, no. 3 (October 1918): 110.

enough money to mark off each brick in red. Brickness, here, is directly linked to the need for a fireproof building, especially as the text refers readers young and old back to Doctor Grenfell's Log in the same issue, in which the near devastating fire at the St. Anthony orphanage is described. It is about collective, distributed, and manageable effort that the brick as both symbol and unit of measure can achieve.

> If the boys and girls will add the number of bricks printed in the various columns and multiply the total by 25 cents, the cost of a single brick, they will see that we have now enough money to buy 16,669 bricks. Dr. Grenfell says he thinks he can put up a brick building for about $40,000.
>
> We have now over one-tenth of the required sum. In the next six months let us double our efforts so that in April we may present Dr. Grenfell with the full amount of money for the new building. Will you do it, Grenfell Leaguers and friends?[74]

Collective action, shared will, creative fundraising, and monetized play all combine to get those bricks moving toward Labrador. The "tale of bricks" implicates readers as donors across ages, and that in turn makes for both the effective, as well as affective, distribution of philanthropic concern over the physical extent of the plant. In order to circumscribe the risk of aged infrastructure, the mission's articulation of infrastructuralism made affective ties between donor and receiver. It makes of Grenfell Leaguers and Grenfell Association members from the United States, Canada, Great Britain, and Ireland into plant workers of a different kind. By distributing the financial burden of a major infrastructural intervention into northern Newfoundland and Labrador, the IGA's fundraising strategy enacts Grenfell's moral ideal of situated action at the smallest scale.

> A large part came in very small sums, many thousands of "bricks," a small pamphlet in the shape of that humble unit of construction, were sold through friends at a "quarter" apiece; and when the sum was completed, and the building finished, from the first sod to the last tile, the whole just shouted once more the absolute value of the very humblest human help, and a new realization that the parable of the widow's mite is only a bit of absolute reality, if we never know anything else on this side of the Great Divide.[75]

This shifting of effective charitableness onto a broad and inclusive settler population was indeed a means of creating affective ties between donors and not only the mission's work but also the materials of the mission's works—its hospitals and hospital boats, things able to be assembled and built. The mission's early evangelical Protestant religious affect became articulated here as an infrastructural ministry that could account moral responsibility for each "unit of construction." It was an effort on the mission's part, as noted above, to have a "collective affect" inhere in its infrastructural repair work and give it "its inventiveness and promise of dynamic reciprocity."[76] While this is a common and ongoing way in which to connect donors to causes, it was particular here in that the bricks were meant to embody, through precise calculation (how many bricks go into making an orphanage?), individual and calculable Christian good works, "the very humblest human help." A tally for the mission's infrastructure of care and repair.

It was also a way in which to expose the mission's donors to the true cost of the enterprise. With only the most nominal sums coming from the colonial government, the mission was run almost exclusively on donations, whether from brick capital campaigns, Grenfell's annual lecture tours, or, as the mission's volunteer and workers without pay alumni base grew (especially among the Ivy League graduates of Yale and Harvard), endowments. It was an analogue awareness to the one Grenfell was also seeking to reveal in his books. As the text on the book jacket of *Down to the Sea*, taken from a *Chicago Evening Post* review, asks: "'Do you know the price of fish? Do you know, that is, the factors that make up the real cost—the exposure, the loss of life, the bitter effort, the heartaches? All these must be included. If you never have totalled the cost, then Dr. Grenfell's new Labrador stories will seem like fiction to you; but they are fact."[77] In this calculation, donating a brick would seem like balancing the material and moral books. And it was that exposition of fact that bolstered the mission's finances. Grenfell's early lecture tours took him through Canada and the United Kingdom and soon expanded into the United States, more precisely its northeast coast in New England and New York, where Grenfell received the most generous (that is, financially lucrative and generative of interest) welcome. Within the span of a few years following his inaugural tour of 1893, Grenfell had secured a definitive place in the closed world of "Fifth Avenue Liberalism": bankers, lawyers, college presidents, and representatives of such financial interests as the Morgans or the Vanderbilts all came to support, via administration and financial contributions, Grenfell's mission work.[78]

It was not only this Christian worldview that allowed for the "Tale of Bricks" fundraising effort to be effective. Another factor was the changing public perception of medical care itself. While social welfare states and publicly mandated and funded health care, in the North American context, were still a number of years away, socially perceived advances in preventative medicine were increasingly circulating across a broader spectrum of social classes.[79] While so-called paupers in the colony of Newfoundland had basic provisions made for their medical needs, it was crucial that the fisherfolk of Labrador were a (hard) laboring population, ill-treated "Vikings of Today"[80] who deserved medical care in spite of their remoteness and because they were out on the water doing the dangerous work of provisioning the empire's economy with income derived from fish, if not the fish themselves. As John Hornsby, coauthor of the book *The Modern Hospital*,[81] remarks in the October 1915 issue of *Among the Deep Sea Fishers*: "The hospital of today is not a thing within four walls; it is not a house that shelters the sick; it is a state of mind. It permeates the whole community; out from its walls radiate all its influences. Like educational institutions everywhere, it teaches and trains those who are to minister, and these in their turn go out into the community and carry its blessings."[82] This medicalization of the environment, and even the individual's cognitive approach to that environment, serves as a good characterization of the socially responsive mediation underlying the mission's concern for its open-ended infrastructural plant. Philanthropic action had to become a state of mind; the bricks of buildings had to be made to speak to a donating public across its full demographic spectrum. How to make them speak relied on a process of collective education that was similar to that of Hornsby's vision of community medicine. In order for infrastructural work to matter to Grenfell Mission supporters, they had to internalize it as a practice that required both moral action and material upkeep, an evolving cycle of progress leading to permanence and increased durability: a ministry of infrastructural care in the service of sustaining an Anglo-Saxon settler colony that was part of the origin story of Britain's imperial realm.

The question of shifting standards returns here. It is worthwhile to note that the mission's 1927 hospital in St. Anthony was the first building for which a loan was required to begin construction.[83] This fireproof, concrete modern hospital was an aspiration that exceeded the donating public's capacity to produce bricks. It also reflected another shift in standards, from wood to fireproof brick and concrete blocks, which disrupts the once commonly held assumption that vernacular building forms emerge from environmental necessity.[84] In addition, it marked an evolution in the infrastructural work that the resource frontier began to necessitate given that, by the 1920s, the mission's donating public was increasingly interested in permanence of care and institutional longevity in northern Newfoundland and Labrador. While

FIGURE 1.7. Emma Demarest, "The Enlarged Hospital," in Demarest, "History of St. Anthony Hospital," *Among the Deep Sea Fishers* 23, no. 3 (October 1925): 104.

it was a settler colonial history that they were nearsightedly trying to shore up, they nonetheless saw the colony of Newfoundland in its broader narrative history as the first territory in the New World to be claimed by the English and yet the last to be granted the formal, imperial status of a permanent colony. Its "retarded colonization" and being subjected to various policies that made it into an economic transit zone gave it that tainted status of transient ship rather than developable shore. As such, the mission, working across shifting standards, sought, in a sense, to recolonize northern Newfoundland and Labrador, to treat it as an environmental medium that could be brought up to a metropolitan and synchronous standard of living by designing, building, and repairing a necessarily ever-expanding physical plant. Its Protestant homiletics included infrastructural work as a cyclical means of sustaining the ongoing process of mediation that the colonial resource frontier made apparent. Yet the mission plant, and the second wave of missionary colonization at its root, also contained the collective affects of a donating public that participated in bringing the fisherfolk back up to speed and into the times and standards of their progressive age. It was a collective project, a network of philanthropic concern localizable on a map or in a brick building. Grenfell would continuously seek to shore the hospital up as a scalable moral endeavor, one that could be "yours:"

FIGURE 1.8. Charles S. Curtis Memorial Hospital outbuilding, St. Anthony, Newfoundland and Labrador, 2011. Photograph by author.

Won't you help with a brick, for a corner, or enough bricks for a buttress, or enough to tile the floor of the laboratory, or to lay the perfect surface of the operating room, or to build the open fireplace for the convalescents' hall, or to put in the back door or even the front one, that will welcome the tide of our suffering fellow creatures, who, in the years to come, will be seeking through it the only help that stands within their reach—and may one day rise to call you blessed?[85]

At the close of World War II the new superintendent of the mission, the aforementioned Charles Curtis, oversaw a comprehensive plan to build "a new hospital and doctor's residence at Harrington Harbour; nursing stations at Forteau and Flower's Cove; a reservoir, dam, power plant, and tuberculosis sanatorium at St. Anthony; and a new hospital at North West River, as well as other improvements to follow."[86] While largely funded through an endowment established by Miss Louie Hall of Rochester, New York, this continuation of the evolving plant also signaled a shift away from private philanthropy toward governmental finance, with the colony of Newfoundland joining the Dominion of Canada in 1949. "Movement is what distinguishes infrastructures from institutions," Berlant writes.[87] The infrastructural mediation the Grenfell Mission fostered over the course of the twentieth century was precisely a constantly renewing relation that responded to the changing international economy of the fishery (as I address in chapter 3) and the mission's prioritization of settler livelihoods over long-standing forms of Indigenous precedent. Fisherfolk lives were part of the environmental medium of the resource frontier. The mission wasn't so much a stable institution that sought to "enclose and congeal power,"[88] but rather it undertook a mobile and shifting form of infrastructural work-as-repair that attended to the becoming, to the constantly mediated settler life of its plant: an infrastructural homiletics of planting.

* * *

SLOW DISTURBANCE, "5 CANADA"

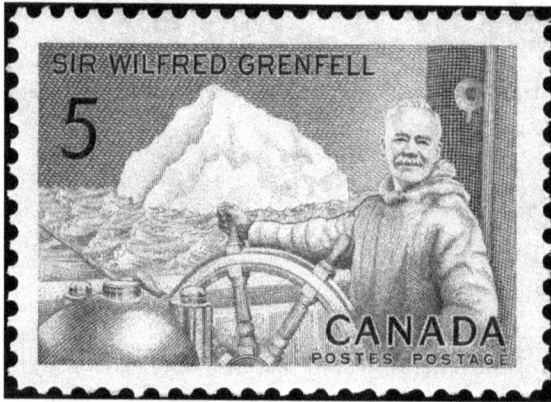

Framed Canada Post Grenfell Mission commemorative stamp, hung in a corridor adjacent to the Curtis Memorial Hospital Rotunda, St. Anthony, Newfoundland and Labrador, 2011. Photograph by author.

2. CREDIT AND COMMON SENSE

"Do you know what used to be wrong with us?" my friend asks. "We were looking always in the wrong direction. Either we looked up to the sky for things that never came from there, or we looked down expecting to pick them up at our feet. Both were wrong. We never looked around us—and so missed the most important thing of all. For only when you look around you do you see the man you stand next to. What one cannot do, ten may." —I. NEWELL, "Credit and Common Sense," *Among the Deep Sea Fishers* (1941)

In the photograph, seventeen men sit in two rows. All wear hats of a surprising variety. No two seem to be alike. Above them, in white block letters, reads RED BAY COOPERATIVE STORE 1896. Written on the side of the wood board structure, with nearly every letter and number taking up its own plank, the sign seems temporary, almost as though it was not meant to last. In the bottom right-hand corner, someone has written on the photograph in black ink. It is hard to discern because the writing appears on the bottom, or top, of a barrel turned on its side. The cursive handwriting, a clue to its authorship, spells out "Cooperative Sign Red Bay." In the center of the front row, at the very center of the photograph, a younger man sits with a paper document displayed in his lap. Directly behind him sits Wilfred Grenfell, slightly askance, bearing witness to what in his estimation would be a new era of cooperative finance in northern Newfoundland and Labrador.

It is a remarkable document because of the inherent locally lived rules that the fishermen were breaking in order to establish a system of cooperative purchasing and selling. As Grenfell tells it, the sign is in chalk and written

FIGURE 2.1. "First Cooperative Store Property of Dr. Grenfell Please Return." MS 254, Wilfred Thomason Grenfell Papers, Manuscripts and Archives, Yale University Library.

on the side rather than the front of the building because the fishermen were wary of upsetting their lines of credit with the local merchant. They wanted the warehouse to be inconspicuous, the sign erasable, and the ability to go back to their previous conditions, known as the truck system, which had been poorly livable, but livable nonetheless. If the gospel of cooperation that Grenfell was preaching didn't work, if it failed like so many fishing seasons of the past, and if they had to revert to what was essentially a barter system without cash that made of them indentured laborers to the local merchant and the networks of his firm in St. John's, London, and beyond, in that case they certainly didn't want such documentary proof of their betrayal. For Grenfell, it was all about the Red Bay Cooperative Store, the mission's first, as a definite sign that cooperation had arrived on the coasts, albeit one of chalk and emulsion.

The Grenfell Mission is usually portrayed as a medical mission above all else. This is, in part, a marker of the material, largely medical infrastructure that has been its evolving legacy over the course of the twentieth century and one that I touched on in the previous chapter. It is also an instructive bias to recall in determining the cultural, political, and social boundaries of an organization such as the Grenfell Mission—the real-world materiality of infrastructures that are appropriable, insurable, and easily repurposed. However, the evolution and transmission of many of the mission's crucial reformative practices that were part of its project of infrastructural mediation are more difficult to track, their signs harder to make out. *Cooperation*, as perhaps the central vanguard ideology that Grenfell, as superintendent of the mission, held to, is what I examine here. It was not only made manifest across the material registers of the mission, in ledgers, scarce cash, quintals of fish, and the photograph that opens this chapter, but it really came to shape the settler's extractive environment, which I treat as a medium in its own right and that served to further the mission's aim of sustaining northern Newfoundland and Labrador as imperial fishing grounds through its wider practice of infrastructural mediation.

As Peters notes, extending Innis's insights around the development of staple economies, "each new medium breeds a cadre of specialists who figure out how to manipulate and program its special carrying capacities and standards."[1] The resource frontier the mission was mediating into being through its infrastructural work was a settler medium that reflected a long and drawn out, equally spatial as temporal, process of dispossession. This was part of the mission's projected reforms on these coasts—to reshape and extend the fisherfolk's extraction-reliant financial lives by educating them through the tenets of cooperative action. The colonial political economy that Innis critiqued, particularly its continual marginalization of, here, the on-the-water producers of wealth, was one that the mission was trying to recast through its commitment to infrastructuralism[2]—its Protestant logistics of care bound to a conception of their extractive environment as a medium that intersected with that of money and its manifestation in the truck system.

Cooperation really crosses the mission story from first to last, up until Grenfell's death in 1940, and even into the mission's activities of the 1960s. Cooperative efforts made by the mission were almost concurrent with the establishment of their first hospitals and nursing stations. On those earliest voyages of the *Albert* in 1892 and 1893, it was clear to Grenfell that the prevailing conditions of the outport communities and among the itinerant fishermen and their families undertaking the late summer Labrador fishery were not merely the result of environmental determinism (harsh weather, poor diet, for settlers an unforgiving, nonsubsistence nature) but also structured by an economic system that ensured that the fisherfolk stayed indebted and impoverished and so visibly and medically ill. Both conditions, medical and economic, had to change in tandem.

In this forced evolution of financial systems, debt held a strategic place for both fishermen and merchants. The political economy of the North Atlantic fishery relied on debt to make its tenuous frontier resource extraction viable, and so creditors and debtors became part of the financial ecosystem of the region.[3] In order for the limited, already marginalized, immigrant and colonial labor pool to obtain the necessary equipment and supplies in remote areas with only seasonal access, capital investment was necessary to get the industry on its feet. From a merchant firm in Jersey, off the coast of Normandy, to its subsidiary in St. John's to its local merchant on the coast of Labrador, capitalized goods would go one way and, usually, quintals and quintals of cod would go the other. With all these middlemen between the extraction of the resource and its ultimate end-point markets, fishermen were beholden to the price their local merchants set for their fish. Grenfell wanted to break this dependence and introduce a cash economy. It was another case of importing an established, if still marginal, British practice, coordinated wholesale purchasing and selling, and establishing it on the outer reaches of the empire—infrastructural mediation had to emerge through the reshaping and projection of a reconceived settler economy. It was the reclaiming of the "principle of mutual effort" as a guiding economic rationale, with that principle itself taking ambiguous root in the mission's religious context that asked of individuals, under free market conditions, to love their neighbors as they would love themselves.[4]

From the 1890s to the 1930s Grenfell and the mission sought to bring the epicenter of the British cooperative movement, Manchester, England, and the Co-operative Wholesale Society, with its attendant quasi-missionary apparatus of hands-on instruction, pamphlets, and lecture tours, to the coasts of northern Newfoundland and Labrador. As part of his religious worldview of Christocentric worldly action, Grenfell sought to bring the entire infrastructural apparatus of cooperative finance and material currency up against the prevailing structural conditions of debt and credit of the truck system. Yet what place did cooperative finance as an environment-dependent practice hold in the mission's activities? Grenfell's economic reform activities often placed him at odds with IGA administrators, as they were seen as taking away from the mission's main purpose of providing medical aid. Also, his efforts to restructure the colonial fishery touched on deep-seated cultural understandings of debt as an accepted and acceptable way of life in northern Newfoundland and Labrador. Debt was, in place of currency, the medium of exchange that bound together social conditions in the colony. As problematic as this was as the basis for a cohesive class identity among a notoriously individualistic fisherfolk population, they were bound by this medium of exchange and by the book debt that left them in a common if detrimental cause that barred them from getting a better market-set price for their fish, cash in return, and noninflated prices for necessary supplies and equipment. In this way, the truck system for the fisherfolk in the colony became a form of "existential debt" that could co-opt not only their laboring lives but also their actual and social lives.[5] The mission viewed debt as an existential condition that could be remedied through infrastructural work—sermons advocating economic justice, equitable accounting practices, and fisherfolk-owned and -operated ships.

In this chapter I examine the emergence of debt-driven outport capitalism and the ways in which it influenced the merchant-fishermen's social and economic relationships prior and up to Grenfell's early cooperative efforts in the 1890s. I unravel how cooperation was consolidated through the mission's gradual practices of infrastructuralism that attempted to reshape the creditor-debtor relation, the environment-dependent medium of money, and the diverse means of rent extraction performed by local merchants and colonial firms that bound that relationship together. I also relate the early history of the Red Bay Cooperative Store in more detail in order to show how the truck system worked on the ground. Finally, as the mission's understanding of cooperative action developed, its tenets shifted as it sought to take stock of its modest material successes and, on Grenfell's urging in particular, to firmly embed the "spirit of cooperation" in the cultural politics of the fisherfolk's future economic practices. The mission, at this time, began to understand the fishery undertaken by the "Vikings of the North" as being reliant on the inevitable granting of credit. Its practices of mediation thus shifted to include a medically anchored understanding of infrastructural care that had to extend to practices of proto-banking.

Outport Capitalism

One year after Grenfell's death in 1940, I. Newell extolled the benefits of cooperative finance in *Among the Deep Sea Fishers*. He opens his short article by setting the scene in Someville, Newfoundland, population 116, in June. He is standing in the doorway of the outport's cooperative store along with its manager, a "short, deep-browed man with arms like knots of steel." In reflective moods, they smoke their pipes and watch the sun going down over the western horizon. The manager, Newell notes, is a man of "larnin" for these parts, having completed the sixth grade before heading to the Labrador fishery. He is also indicative, for Newell, of the success of cooperative education in Newfoundland and Labrador. "The Co-operative movement, without fanaticism, without sentimentalism," Newell writes, "calls on men to pool their united efforts in a common-sense way that works. It puts idealism to work, and pays dividends in fellowship." Newell's article marks the opening of the first Co-operative Credit Society in northern Newfoundland, located in St. Anthony. It is a watershed moment in the historic creditor-debtor relationship between merchants and fishermen in the colony.[6]

Following the article is an excerpt from the St. John's *Daily News* of June 4, 1941, announcing the formation of the Co-operative Credit Society, "an important factor to the people of St. Anthony in their struggle to have complete economic independence." The *Daily News* article goes on to describe Grenfell as "the first co-operative field worker in the area."[7] It is worth keeping in mind how economic reform undergirded so many of Grenfell's and the IGA's schemes of reform. By 1941 the missionary of memory had become, at least in part, a cooperative field-worker and educator. The much-vaunted and highly sought-after independence of the fisherfolk had to be economic first. The mission worked toward that goal by combining the pragmatics of cooperative education with the scriptural, spiritual, and definitively infrastructural work of the social gospel.[8]

Just below the newspaper clipping inserted into the article is another excerpt from Grenfell's *Labrador Logbook*, a collection of short glosses on keywords, events, and important figures. Under the heading "Samaritans," Grenfell specifies that "in the civilization that is coming, love for one's neighbour in every department of life will call for cooperation as the only basis which can ever be permanent."[9] These various positions on fellowship, the manager's horizontal equality and recognition of collective power, Newell's commonsensical pooling of resources, and Grenfell's making of an all-encompassing neighborly love the basis of exchange combine to sketch out the progressive place that cooperative finance could hold in the increasingly politicized imaginaries of the fisherfolk of northern Newfoundland and Labrador in the early 1940s. With more concerted efforts at unionization and collective organization having taken shape farther to the south and earlier in the century, most notably the Fisherman's Protective Union (FPU) under the founding influence of William Coaker, it is remarkable that cooperative finance was relatively slow to take shape and hold in northern Newfoundland and Labrador.[10] While this was due to prevailing social and economic conditions that included a near absence of educational, medical, and judicial institutions in the colony's northern reaches, it also shows how difficult it was to keep the North Atlantic fishery profitable across its scales of production, from merchant down to fisherman and back up again.

Grenfell's efforts in the 1890s came on the cusp of the fishery's shift from merchant to industrial capitalism. While fisheries in Iceland and Norway were relatively quick to adopt novel fishing technologies, incorporating new refrigeration techniques, steam-powered, larger, and more efficient trawlers that needed smaller crews, merchant firms in Newfoundland and Labrador lagged behind, and the colony's gradual slide into bankruptcy by the 1930s was in part the result of this reticence on the part of merchant owners in St. John's and London to make the necessary capital-investments to compete in international fish markets. Add to this, over the course of the 1800s, an economy that was ever more both capital- and cash-scarce and also increasingly subject to international pressures, and credit and debit become fundamental mechanisms for the production of, and production within, Newfoundland and Labrador's staple economy. Under so-called difficult market conditions, with an uncertain source of production, a labor pool of morphing size and reliability (through illness, bankruptcy, emigration), and a commodity of poor quality and price when compared with international competitors, debt was the interface that enabled the perpetuation of the colony's economy. Merchant capital, loaned to independent, small-scale producers, sought to make individual enterprise a reliable part of that economy.[11] In this environmentally determined context, it was debt rather than hard currency that functioned as the primary medium of exchange.

For merchants and fishermen alike, the frontier economies of the Grand Banks and Labrador represented tremendous rewards and risks. The former involved organizing a system of production in remote areas with seasonal access, open to limited competition, with a severely limited supply of specie and with the double-edged sword of a dependent labor pool comprising both producers and (indebted) consumers. For the latter, one of the few ways to enter into the costly frontier economy was to obtain from the merchants, on credit, all of the necessary equipment and supplies to ensure their own survival and productivity.

One of the pillars of this economy, at the intersection of the interests of merchants and fishermen, was the truck system. G. W. Hilton defines truck as "a set of closely related arrangements whereby some form of consumption is tied to the employment contract."[12] Employers, for instance, could pay laborers in goods rather than wages, issue specie valid only at the company-owned store, or advance goods to laborers in place of wages. This system was rightly criticized for tying laborers to employers through the device of debt as well as for creating a monopsonistic dependence that could only be played out, usually in the form of generational debt, in the ledger books of the local merchant. However, as Rosemary Ommer notes, it was also a way to compensate for the lack of adequate financial infrastructures in colonial outports.[13]

While I will later touch on the cultural dimensions of debt, it is worthwhile to note here that Grenfell largely did not take into account the multi-scalar, transnational problem of the creditor-debtor relationship when undertaking his early cooperative reforms. This is not to claim that he was not aware of the complexities of the international fishery or of the mission's own place within the political economy of the colony and the empire. Rather, it is to see Grenfell's motivation as, primarily, a combination of religious pragmatism as expressed through social gospel and financial justice. As with the structure of so much of his writing, scriptural phrases and events provided the interpretive base from which worldly action could, and should, lead.

In the chapter titled "The Future of the Mission" in Grenfell's 1919 autobiography, A Labrador Doctor, he looks toward the ongoing development of the mission's activities while also cautioning that "the Mission works should one day become self-eliminating."[14] For Grenfell, cooperation was the path toward the sort of sustainable self-sufficiency that would one day render the strictly charitable work of the mission unnecessary. He was not content to merely follow established orthodoxies; he believed that the "Christian choice is that of Achilles." By this he means that we should question the foundational principles of living, that "Nature also teaches us that the paths of progress are marked by the discarded relics of what once were her corner-stones."[15] Keeping his own advice in mind, Grenfell diagnoses the current state of the North Atlantic fishery and the resources available to it. Deeming it able to sustain its current population of resident laborers and their families, he cautions that this state of affairs will be brought forward only "when science and capital are introduced here, combined with an educated manhood fired by the spirit of cooperation."[16] In this equation, "science" and "capital" are the static, material elements that can be put into place, whereas "spirit of cooperation" is a more nebulous force: emotive, stemming from a shared religiosity, and, most likely, more difficult to come by and sustain. Unlike the mission's hospitals and nursing stations, cooperation was a movement that could encompass laborers' lives and one of the few opportunities afforded to the fisherfolk to actively participate in the improvement of their material conditions. As such, cooperation grew out of the existential and environmental conditions the fisherfolk inhabited and, for the mission, became a practice that could project a reformed "mode of being."[17]

Reconceiving of debt was part of this new existential horizon, and it highlights how the mission's practices of infrastructural mediation were bound together through their frontier environment as an expanded medium that could make manifest their settler infrastructural work. That the mission was seeking to recast the existential stakes of this North Atlantic resource frontier indeed shifts an environmentally inflected understanding of debt toward the expanded medium concept; as Peters highlights, "Once communication is understood not only as sending messages—certainly an essential function—but also as providing conditions for existence, media cease to be only studios and stations, messages and channels, and become infrastructures and forms of life."[18] Following the nineteenth century folding together of transportation and informational exchange as instances of "communication," the mission was attempting to devise the right medium that could sustain a form of equitable extractive communication for the fisherfolk of northern Newfoundland and Labrador.[19] It was a case of projecting these "infrastructures and forms of life" into being and having them register across a materially reformed (and thus communicative) medium-resource frontier. Settler colonialism in the colony was indeed a practice of ordering that was so tightly bound with its ambient environment that their existential stakes became intertwined. If media are "ensembles of natural element and human crafts," then the mission's reformed resource frontier was a medium caught in the process of settlement—an interaction between habitats and this laboring *anthropos*.[20]

Yet for Grenfell what this material and infrastructural cooperation also ultimately relied on, beyond the productive neglect of the creditor-debtor relation, was neighborly love; "love builds for a future, however remote," Grenfell writes, "and at present we see no other way than to work for it, and know of no better means than to insure the permanency of the hospitals, orphanage, school, and the industrial and cooperative enterprises, thus to hasten, however little, the coming of Christ in Labrador."[21] It was a love that Grenfell partly tried to simply write into being, as so many of his books and tracts interpreted, glossed, and narrativized fisherfolk livelihoods and lives into an existence made up of Good Samaritan care and actions. While conceding that the fishery was a highly individualistic enterprise that pitted fisherman against fisherman, he believed that the mission's works—its material infra-structure—could only endure by propagating a constructive love through a common creed that was fundamentally that of Christianity at large: "a new possibility of cooperation between God and man through the coming of Christ to earth."[22] This is what infuses that "spirit of cooperation" that Gren-fell saw as crucial to scientific innovation and capital investment. Specifically, the coming of Christ to Labrador meant the reform of a substantially aged and flawed merchant outport capitalism that relied on a morally and mate-rially suspect creditor-debtor relationship. Cooperation was to be built as an affective religious bond that could circumvent the pressures of outport cap-italism and begin to build out its own infrastructural, "spirited" manifesta-tions of success.

In a 1938 letter from Grenfell to Cecil Ashdown (by then, executive director of the IGA) written two years prior to Grenfell's death, he notes that his "own reason for going to Labrador, or among fishermen anywhere, was primarily because I considered the world owed fishermen a debt since the time of the fishermen disciples and what they had to bear to incur it."[23] A moral debt, for Grenfell, became a doctrinal creed of love that would ultimately try to redress the inequalities within the colony's economy. Grenfell's mission, his drive for reform, was shared by several Protestant denominations in the early twenti-eth century. In the United States, the Presbyterian pastor Warren H. Wilson began "community study" initiatives that sought to bring such issues as "or-ganized labor, Socialism, child labor, women in industry, class consciousness, social and economic problems, industrial education, housing, sanitation, un-employment, etc.," under the influence of a newly "insurgent Sunday-school," whose pupils would go out into the community to assess economic problems, demography, poverty, and class distinctions, among other issues. Seeking to pair up with "social welfare movements, public officials, and educational bodies" to document and improve conditions in a specific church's surround-ing area, religious education, especially in the Episcopal Church, became an active, quasi-sociological process of observation, assessment, and action.[24] For Grenfell, this documentary work would have been part of the broader commitment to the social gospel and, combined with the sound economics of the cooperative movement in England, could make for a concrete outlet for his evangelical Protestantism derived from Dwight Moody, the American preacher who was instrumental in Grenfell's heeding the Christly call in the 1880s.

Cooperative action thus became inextricable from the mission's broader work of infrastructural mediation. As Grenfell remarks, "The cooperative movement is the same question [as that of education and modern pedagogy] seen from another angle, and is almost contemporaneous with our earliest hospitals."[25] It was a question requiring a doctrinal as well as material response, a "debt" that had to be repaid through good works in its double sense, with religious "spirit" bound together with an infrastructural form of material care. If it was cooperation as a "sign" that was to precede both the chalk and the emulsion, it would have to grapple with the creditor-debtor relationship on that same plane of neighborly love. Just as the mission's infrastructural care and repair work relied on its integration into its conception of the Protestant homiletic tradition, so too did cooperation have to be bound together as an infrastructure *and* a form of life. This was a now-historical process that still holds value for scholars in environmental media studies and its allied fields, as it articulates and lays bare how an expanded, environmentally inflected medium came into being: northern Newfoundland and Labrador as a sited imperial entity had to be made to speak through its reformed, Protestant evangelical environment. It was the mission's practices of infrastructural mediation that had to account for this binding of religious affect and reformed financial infrastructure.

The Red Bay Cooperative Store

The fishing village of Red Bay is located on the southeast coastline of Labrador. Looking over the Straits of Belle Isle, it is equidistant from the North Coast of Labrador and western Newfoundland. In the time of the Grenfell Mission, Red Bay was like many other outport communities. It had a relatively small but stable settler population, a local merchant store that was sometimes open only during the fishing season (roughly from May until October), and intermittent, seasonal access by boat from the island of Newfoundland and by dogsled from surrounding communities in Labrador. As Grenfell tells it in his autobiography, the genesis of the cooperative store began when, on a return trip down the coast in the fall, he came on a group of fishermen waiting on their stages to be removed from the coast as they deemed their situation unlivable. "A meeting was called that night to consider the problem," Grenfell relates, "and it was decided that the people must try to be their own merchants, accepting the risks and sharing the profits."[26] He specifies that both the fisherman's and the trapper's life is a gamble, and so it is natural that they prefer credit advances, "for it makes the other man carry the risk."[27] For the merchants, potentially unreasonable risks came with consistently unreasonable rewards. If the fisherfolk were to become their own merchants, they had to be willing to run their own risk, both in extraction-as-production in the seasonal fisheries and on international fish markets.

Francis Hopwood's rallying letter to the MDSF to bring medical aid to the settler Anglo-Saxon population documents many of the unjust economic and cultural relationships prevailing between merchants and fishermen. He decries the continued existence of the truck system, "so absolutely condemned [in England], [and that] has not yet been destroyed in the Colony," and notes that the interests of merchants and fishermen do not coincide.[28] Moreover, with the vast majority of the merchant class making up the governing body in the House of Assembly, he remarks how there is very little hope that the interests of the fishermen will be fairly represented or that their living conditions will change. "The fishermen, who represent almost the entire labouring population of the country," Hopwood writes, "are in fact dependent on the merchants for the very necessaries of life; it may almost be said that the merchants own the fishermen."[29] This recalcitrant acknowledgment, like Grenfell's, of intra-Anglo-Saxon slavery, is meant to be the ultimate spur to the MDSF to act quickly and efficiently as well as beyond its usual sphere of medical aid and evangelical spiritual counsel. In his concluding remarks, Hopwood uses the racial bedrock of the fisherfolk as a form of leverage to induce change in the colony: "I would, finally, point out that, in making this new departure and entering upon a fresh field of labour, we are not interesting ourselves in people who do not want help, and who have nothing in common with us. Complaints have often been made that philanthropic people are disposed to subscribe to apply pocketandhandkerchiefs [sic] to little [racial expletive]s who did not use them; that work is undertaken abroad whilst so much remains untouched at home."[30]

Hopwood specifies that these Newfoundlanders are "British subjects" who have "votes." As such, they are a population that want help and are "men [who] are capable of education and enlightenment in every way"; that is, they are racially receptive and available Anglo-Saxons who can be newly inculcated with the values of the "Old Country," "which must have a permanent effect for good upon their social and political conduct."[31] While these are relatively common late Victorian mores, albeit charitable ones, they show what sort of reformative practices the MDSF and Grenfell, in the early 1890s, thought the colonial fisherfolk warranted. It was as if they had a duty to bring these colonial British fisherfolk up to the "Old Country's" moral, material, and laboring standards. While these fisherfolk operated on the margin of the empire and on the margin of an international economy that was often poorly regulated, they could still be made into an economically productive and morally worth-while populace.

For Hopwood, "in these days of projected confederation and active inter-est in the Colonies,"[32] it was clear that in outport communities such as Red Bay the colonial settler social situation could be remade along the best, most progressive of British lines. Grenfell would continually recast this under-standing of "standards" and "improvement," yet the foundations of Anglo-Saxonness as a reliable and reformable racial populace would remain, often in a background of common cultural assumption in many of the mission's prac-tices of infrastructural mediation. In the process, the settler colonial politics at work across this labor force essentially left out the Indigenous residents of Labrador—considered as "races" apart with their own environmental deter-minisms and hereditary traits to contend with. The settler colonial lives the mission's practices of reform were enabling were thus racially and environ-mentally circumscribed. Its imperial and evangelical Protestant worldview, exemplified through the creation of the Red Bay Cooperative Store, held that infrastructural mediation grew out of the real-world conditions of Anglo-Saxon territoriality and empire building. The extraction of fish was a practice that circled back to inflect the set of relations that the mission's human actors established with their ambient environment—like a wharf built atop a rocky shore, a settler resource frontier that was a medium made to register the en-during ambitions of Anglo-Saxon racial superiority.

Grenfell would tell and retell the story of the creation of the Red Bay Coop-erative Store. The most detailed narrative comes in the 1929 pamphlet "Lab-rador's Fight for Economic Freedom," the nineteenth installment in the "Self and Society Booklets" series published by the liberal publisher and popu-lar economist Ernest Benn and printed and produced by the Co-operative Wholesale Society of Manchester. The thirty-three-page pamphlet is divided into six sections. The first, "Self and the Other Fellow," reinforces Grenfell's belief that "[c]o-operation needs love for the other fellow."[33] However, he complicates this "love" between human fellows by introducing a degree of social Darwinism.[34] While Grenfell acknowledges that "our people in Lab-rador and North Newfoundland are good, average specimens of the *genus homo*," he wants his readers to remember that "permanent evolution is al-ways slow," that "the primal instinct of self-preservation from caveman on-ward makes human beings conservative in proportion to the similarity of their environment to that of the primitive era."[35] This cautionary caveat that essentially says "they may be Anglo-Saxons, but they live on the harsh, unforgiving, and formative coasts of northern Newfoundland and Labrador, and so will inevitably evolve at a slower pace than the rest of civilized, metropolitan humanity" is meant to account for the continued difficulties the mission's introduction of the cooperative movement faced, most notably, that of the individualism or "self-preservation" of the fisherfolk. As a result, the mission's understanding of an infrastructural religious affect included these sorts of ties to the environmental conditions of its North Atlantic resource frontier—it was indeed as a "frontier" rather than a preexisting ecological zone that the mission practiced its sphere of operations. The expanded medium concept, one that accounts for the capacity of environments to signify, moves out toward the infrastructural mediation that the mission created: they speak together and tell how environmental conditions could be co-opted to denote racialized modes of communication.

Despite these new contextual markers (pace Herbert Spencer), in Grenfell's framing of cooperative action as expressed in this pamphlet is an effective narrative of the development of the cooperative movement in northern Newfoundland and Labrador.[36] In the second section, "The Slavery of a Truck System," Grenfell holds nothing back in denouncing the injustices of the truck system and links those injustices to the seventeenth-century practices of the Hudson's Bay Company. With essentially no currency in the colony, its Indigenous inhabitants undertook a barter system with the corporate entity that, over the course of centuries and various phases of evolution, remained relatively unchanged. For the Indigenous peoples of Labrador, this partly meant an increasing marginalization in the northern fur trade and a silent, technologized encroachment, via barter commodities received as payment, into their established living practices.[37] Moreover, the colony's laboring populations shifted to be made up of early emigrants, from the seventeenth century forward, from the west coasts of England and Ireland. By the nineteenth century they were multigenerational settlers of outport communities and, to some extent, resident colonials. Along with population growth came expansion of the fisheries, moving by the late nineteenth century from the Grand Banks to the coasts of Labrador. This expansion of the fish economy, in turn, led to increased provisioning of boats to prosecute the fishery and a concomitant increase in merchant credit as well as bank loans. In 1894 the cyclical exchange between creditors and debtors came to a halt with two out of three of the colony's largest banks, the Union Bank and the Commercial Bank, both privately owned and largely managed by merchant fish interests, collapsing under impossible debt loads. In an already scarce cash economy, the closure of the Union and the Commercial as currency minting institutions meant that for remote fishermen and their families the prospect of an intensified truck system was a real possibility.

The Red Bay Cooperative Store was launched in the fall of 1896. With an original group of twenty-five families, "the original members all being Englishmen by birth,"[38] as Grenfell specifies, only one man could read, write, and keep accounts at a sufficient level to take charge of the store. The first difficulty was that of carrying over generational debt. In order for a cooperative store to get off the ground, its shareholders had to be virtually debt-free; otherwise, their new assets could be pursued as payment on old and existing debts. Grenfell cites the example of an unnamed community of four hundred and eleven people, where a local trader went into bankruptcy after accumulating a debt of $64,000. Many of these outport communities lived on credit, or book debt, alone, and so the introduction of a circulating currency was viewed with suspicion as a complicating factor in their mode of exchange with local merchants. In Red Bay the first task was to lift the fisherfolk out of debt. The first year their motto was "Let us eat grass for flour and get out of debt."[39] With this goal in mind and with the initial subscription drive complete, the shareholders managed to raise only $85 in initial capital. As a result, Grenfell and the MDSF became a de facto credit-lending institution, extending the necessary capital for the foundation of what had come to be called the Copper Store. At base, the cooperative store was meant to combat the uncertainty and risk associated with the fishery by distributing its costs across the community. Risk was perhaps the defining condition of fisherfolks' existential horizon, and so it was, like debt, a binding part of their "mode of being" under outport capitalism. Common merchant practices consisted of either leaving the outports at the onset of winter, in the process leaving the local population uncertain of their ability to hold over until the following spring, or charging winter prices for their already inflated goods, given that the last steamer of the season had departed and there would be no chance of getting the item by any other means. By making themselves the suppliers of their own goods, the fisherfolk of Red Bay could pay the real costs associated with their geographic position and place in the supply chain.

The mechanics of the selling of the cooperative's fish were both simple and, for the time, trailblazing. Prior to the Red Bay store, the direct sale of fish to markets or individuals for cash was considered a black market of sorts, and the fishermen who engaged in the practice were often disciplined, both judicially and financially, by local merchants. Once the pool of fishermen had determined the size and quality of their seasonal catch of salt-cured fish, a schooner was ordered from farther south to come to Red Bay to carry it to market. The mission loaned the money to hire the schooner, though by 1901 they had raised sufficient funds and interest to commission a purpose-built schooner, *The Cooperator*.[40] The "secretary" of the store, adept at his figures, traveled with the fish to market in St. John's to sell the fish, "for the first time in history" according to Grenfell, for cash.[41] The schooner agent also assisted in the sale. Grenfell notes four economies enabled by this system: the schooner, otherwise traveling north without cargo, could bring all the supplies that the catch, once sold, would be able to purchase; the local merchant's middleman profit was saved; the fishermen loaded and unloaded the schooner at each end themselves, thereby avoiding considerable cost; the final sale price of the catch was invariably higher than that granted in the past by the local merchant. This straightforward direct-to-market system had to overcome the lack of substantial initial capital, a cultural politics of debt and credit fostered by the local merchant as a figurehead of centralized control in St. John's or England, as well as the divisions between the fishermen themselves, as each remuneration was collective.

The Red Bay Cooperative Store would, in time, come to be one of the most solidly established of the ensuing cooperative mission stores. The pamphlet goes on to recount the founding of subsequent stores in Flower's Cove on the Straits of Belle Isle, in West St. Modeste and Cape St. Charles in Labrador, and in Great Brehat and St. Anthony, on the Northern Peninsula of Newfoundland. While I will later return to the founding of the St. Anthony spot-cash store, each store raised questions related to their specific place in the broader system of the fishery, and each store either failed or succeeded due to a combination of such fundamental structural issues as literacy levels, generational debt, sectarianism, and established lending practices that proved too hard to break: those infrastructures and forms of life undergirding the mission's understanding of cooperation as part of its living mediation. The logic of the shareholder with an expectant stake in his or her investment did not completely take hold in the Red Bay Copper Store or in many of the mission's other stores. Don't forget that its founding was, in a way, an anonymous act. "At the time of the formation one very significant fact was that every shareholder insisted that his name must not be registered," Grenfell writes, "for fear some one might find out that he owned cash."[42] In the system as it was at the time, cash could be construed as a liability—a mark of black market guilt, an unwillingness to pay toward one's debt, or a questioning of the prevailing book debt system as a whole. The Red Bay Cooperative Store, at its founding, was meant to be invisible. The fact that it was Grenfell who chalked its identity, that announced its, or his, intent to propagate cooperative distribution, was an ambiguous mark, a cursive, literate signature that was slow to be adopted by the resident fisherfolk. It is a mark that shows how the mission's practices of infrastructural mediation were, in part, top-down projects of reform but also efforts inextricably bound up with the everyday existential conditions of the fisherfolks' lives and, relatedly, the financial, medical, and other infrastructural networks on which they relied. The story of the Red Bay Cooperative Store allows us to see how the mission worked through a form of infrastructural mediation predicated on making economic relations anew and grounding them in a settler colonial resource frontier, a medium of both registration and communication that could reflect both racially sound men and their economic morals.

While there is a dearth of materials relating to the opinions held by the fishermen cooperators themselves, Grenfell's co-opting of cooperation's genesis story does nonetheless make (somewhat instrumental) room for their voices. At an early meeting to gauge the possibility of opening a cooperative store in St. Anthony, Grenfell recalls how the traders spoke at length, debating the various merits of the proposed system. Near the meeting's end, as Grenfell tells it, an old, well-traveled sailor who "was always listened to with reverence as a kind of oracle," rose and said: "All I wants to ask is one question, doctor. . . . If this 'Copper store' buys a barrel of flour in St. John's for five dollars, will it sell down here for fifteen dollars?" And, with that, according to Grenfell, "we brought the meeting to a close."[43]

As noted, Grenfell's championing of cooperative action was freighted with his Christian message of neighborly love. While he recognized the accomplishment of getting a higher price for Red Bay fish than at any point in their history, within his larger worldview he saw "co-operative distribution [as] one interpretation of the Christian religion."[44] Within a fishery market system built on risk, debt, international pricing, an unreliable source of extraction-as-production, and marginal capital investment in novel fishing technologies, the extension of credit, neighborly or otherwise, was an essential element of the fishermen's working lives. This goes back to the importance of the St. Anthony Credit Society in 1940. That the mission, a charitable enterprise with largely medical aims, had to enter the colony's system of finance capital is a testament to the nebulous boundaries operating between creditors and debtors in nineteenth-century Newfoundland and Labrador. What debt did the Red Bay cooperators feel they owed the mission?

In 1905 shares in the Red Bay Cooperative Store were worth $5. By 1918, they were worth $104, having consistently paid a 10 percent dividend. Every family in Red Bay was debt free.[45] "The fisherman's and trapper's life is a gamble," to return to Grenfell's earlier cited claim, "and naturally, therefore, they like credit advances, for it makes the other man carry the risks."[46] In this case, the "other man" was also their doctor and provider of cooperative education and capital. "It has saved the village and the neighbourhood, and still does the trade for the people," Grenfell writes, "while surrounding villages have nearly gone out of existence owing to increased cost of living and the increasing impossibility of existing at all under the old régime."[47] Cooperation carried an existential debt of its own. It may have enabled the survival of otherwise defunct outport communities, but it also signaled this shift toward the mission's understanding of cooperation as part of its project of infrastructural mediation that could lead to a much more forceful form of evolutionary intervention by mission workers into the fisherfolk's laboring lives. Between morals and money, existence and cost of living, Grenfell understood cooperative action to be mediating in a fundamental sense: It was meant to enable the coming of both cash and Christ to Labrador.

Existential Debt

In the colony of Newfoundland and Labrador, from the bank crash of 1894 to the loss of responsible government in 1933, there was, as I have touched on here, a debt economy with the natural resource of salt cod acting as one of its principal environmentally inflected media of exchange.[48] Debt, while counted in cash by local merchants, was largely payable only in fish. In this reading, if it was the debt of fish and the extending of credit for supplies and equipment that upheld social relations in the colony, it would seem that the fishery itself was the only possible staple economy, the only livelihood open to merchants and fishermen alike. You had to fish to live, and as such you lived the fishery.

While the colony was indeed mostly viewed as an immense imperial stage or schooner for the harvesting of fish and seals, it was as if the colony as a whole was living an existential debt that tied it to an increasingly problematic fishery. Tenuous as this form of association between merchants and fishermen must have been, it was nonetheless a bond brokered by common debts. As James Hiller argues, "To fishermen and merchant alike, salt cod was primarily a medium of exchange [rather than a foodstuff]. The interest of the former was to produce sufficient to maintain his credit; the interest of the latter to buy as much as possible at the lowest price to maintain his own payments for imported goods."[49] In such an economic situation, nearly every debtor was also a creditor.

Yet this paradigm of exchange as the basis of social relations has been questioned by critics who highlight the disparities that precede exchange at the levels of both production and wage labor.[50] They posit that there is no equality of exchange underlying social relations. Rather, it is an unequal distribution of debt and credit that structures the social. Debt creation, in this perspective, becomes equally the production of evolving power relations between debtors and creditors and of a debtor subjectivity that economizes morality. The introduction of specie into the relationship, as the mission's cooperative efforts were at pains to do, and of creating stores that would vie with local merchants for local business were viewed as opening up the entire production and consumption systems to competition that ultimately would drive down the price of necessary goods and drive up the sale price of catches. This was the vital form of infrastructural mediation the mission was there to undertake—to introduce such a novel medium of exchange (both literal and existential) into the everyday lives of the fisherfolk. However, in order to accomplish this, the mission also had to traffic in this production of a debtor subjectivity. Like the local merchants themselves, they had to begin to extend credit in order to redirect that debtor subjectivity toward a cooperative end.

With this financial and infrastructural missionary intervention and its modification of prevailing relationships surrounding the fishery, the Grenfell Mission was bound to run up against merchant interests. On July 12, 1917, a group representing eight merchant firms presented a petition to the House of Assembly regarding "certain alleged operations of the International Grenfell Association."[51] Underlying many of the charges—foremost among them that the IGA was operating as a sort of shadow commercial enterprise largely funded by its network of wealthy American philanthropists who subsidized its outport cooperative stores—was the fact that the Grenfell Mission, both its activities and workers, was to a degree exempt from and untainted by the colony's historical existential debt that tied together the livelihoods of local merchants and the fisherfolk. The eight merchant firms wanted their allegations confirmed by a governmental inquiry. While the petition was a reaction to the relative success of the mission's cooperative stores, it also reflected how widespread the mission's activities were becoming in northern Newfoundland and Labrador. The mission, in the eyes of the merchants, was a growing threat; its threatening nature, chiefly financial though also implicitly raising concerns around class, governmental responsibility, and social welfare, was starting to become all too apparent. The merchants claimed that the IGA, through their association with stores on the coast, had ready access to imported goods that were exempt from customs duties as a result of the organization's charitable status. They also accused the IGA of providing American philanthropic capital as credit to its cooperative stores without expectation of interest or dividends. Moreover, this capital was raised, the merchants claimed, by "misrepresentations that this dependency of Newfoundland is largely composed of paupers."[52] In a way, the merchants were accusing the mission of being an extra-colonial actor, one who didn't abide by the established rules between creditor and debtor.

The merchants' petition resulted in the filing of a counter-petition by the IGA. In that petition to the House of Assembly, dated August 16, 1917, the IGA lays out its original formation in 1914, the medical and social purposes that guided its founding, the members of its voluntary board of directors, and, in point eight of the petition, a very direct rebuttal to the merchants' charges: "The Directors regret that the Association has been and is still the object of extraordinary calumny and misrepresentation, emanating largely from certain mercantile firms and persons trading on the Labrador who find that as a result of the social work of the Association's subordinates the people of that region are not so easily exploited by commercial interests as formerly."[53] The IGA counter-petition goes on to welcome the undertaking of an inquiry and specifies that they would like it not only to clear them of their alleged abuse of customs duties but also be broad enough to include "the general question of the merits and values to the public of your petitioners' work."[54] Confident in the justness and legality of their charitable enterprise, the IGA took the 1917 inquiry as an opportunity to receive governmental and, ultimately, popular sanction. Many of the St. John's newspapers, certain church leaders, and other merchant class interests had long accused Grenfell and the IGA of banking on characterizations of the colony as poverty-stricken and both physically and morally destitute. While many of Grenfell's parable-like stories of suffering from the coasts of Newfoundland and Labrador did to some extent do exactly this, they were nonetheless accurate if sentimentalized accounts of conditions in outport communities that were off the map of governmental aid.

The 1917 inquiry, according to Grenfell, would ultimately become "the best advertisement possible,"[55] as it refuted all of the merchants' charges and highlighted the benefits of the IGA's activities, including their cooperative infrastructure building. Yet the mission was wading into the murky financial waters of the colonial fishery, both out of necessity, to intervene effectively to better the living conditions of fisherfolk, and out of a sense of Christian missionary reform and education. Cooperation was the better way, that "message of love."[56] However, Grenfell did indeed recognize that the mission's work was transitory. If ultimately successful, its infrastructural plant and cooperative finance as a form of affective religious bond would be passed on to the people themselves: "I realize that the Mission should not own a cooperative store. A store can't be really cooperative if owned either by a Government or a Mission, or anybody but those who benefit by it. Its ultimate success can only be when cooperators reap all the advantage, both of getting and giving the best service that love for their neighbour dictates."[57]

In order to facilitate this ideal, the mission, in a 1916 revision of its memorandum and articles of association, outlined its scope of operations and the ways in which it could go about fulfilling "the objects" for which the IGA was originally put into place.[58] Listing twenty objects, they run a wide gamut from its primary concern "to visit seamen, fishermen and persons engaged in the fishing industry and fishing and other vessels upon the coasts of Newfoundland and Labrador" to the dissemination of educational literature and the purchasing and managing of land in the colony. The list is comprehensive and indicative of how much the mission was physically "on the ground" in the region. This is made most evident in their stated intentions around the question of mission infrastructure.[59] With drydocks, bridges, reservoirs, and dams (as touched on in the previous chapter), the list is more that of an industrial concern than a medical mission. While to some extent the preparatory language of legal domain, it is also indicative of the sort of plant the mission deemed "necessary for its activities or capable of being conveniently and profitably operated in connection therewith." Given that the IGA's fourth article of association states that "income and property of the Association, whencesoever derived, shall be applied solely towards the promotion of the objects of the Association as set forth in this Memorandum of Association,"[60] it makes of the mission's infrastructure a noncommercial system of distribution meant to facilitate not only the mission's work of circulating medical aid, and so a cyclical exchange of specialized goods and trained people, but also an infrastructure of care that could enable the operation of cooperative distribution; infrastructural repair work and its associated forms of religious affect could thus encompass debtor subjectivities as well.

Objects eleven through sixteen address the mission's financial position.[61] With provisions allowing the mission to issue promissory notes, to borrow and raise money "as the Association may think fit," to purchase and operate any business, and to issue loans, the IGA, as an organization, starts to take on the status of a creditor, albeit one that was not after profit. While these twenty objects were in the realm of legal possibility, the mission did nonetheless, as the Red Bay Cooperative Store shows, take on the role of a medium of capital, if not that of a commercial bank proper. Insert them into the merchant-fisherfolk relationship, as an extra-regional actor of concern, and the picture of clear-cut creditors and debtors starts to blur. The social exchange between merchants and fishermen, an exchange in its most literal sense embodied in the truck system, becomes in the colony one nearly entirely predicated on the issuing and clearing of credit and debt.

The mission's effects on that system of exchange eroded the existing relationship by introducing cooperative purchasing and selling. It was a concerted intervention into the everyday moral economies governing the colony's prevailing understanding of existential debt. The mission was thus the primary institutional force that, in northern Newfoundland and Labrador at least, could reshape local debt-driven settler subjectivities, equally moral and cultural, through the introduction of cash as a medium of exchange (in place of the commodity of fish): it was an existential as well as fiduciary conception of credit that the mission created through the infrastructural mediation of cooperation. Taking a *longue durée* view of the mission's influence, one might be inclined to read this as the first step out of a mercantile, outport capitalism of indebtedness and rent extraction as well as the first step down the road toward market competition and the introduction of industrial capitalism into the Newfoundland and Labrador fishery. Once the mediating factor of money, rather than book debt, separated merchant and fisherman, this also opened up the possibility of fishermen seeking credit directly from commercial banks, implicating a whole new system of control through interest, investment of capital in leased equipment, and other property acquisitions.

While it was a mode of social exchange that needed to be changed, particularly in its relationship to the extraction of surplus value by merchants, the truck system created the conditions of possibility for a subsistence economy in a notoriously unreliable resource. Ommer's analysis of the nineteenth-century truck system in Gaspé shows how it was not uniformly viewed as an exploitative system by the fishermen directly involved. Many of them prospered, some remained indebted for life, yet the real injustice of the system lay not in the extension of credit in goods to the fishermen, but in the premiums charged by merchants, as they reduced the fisherman's real wage considerably. There was a need for credit-granting facilities in these outport communities as well as institutions that could create the necessary, affordable, and reliable transportation and financial linkages with international markets.[62] In the absence of commercial banks being willing or physically able, inasmuch as they were not being represented in outport communities, to issue loans to small-scale fishermen engaged in the fishery, merchants took up their roles as middlemen. The Grenfell Mission tried to intervene in that system by performing as a credit-granting actor of a particular kind, as its credit was both existential and financial—a binding together, a making of infrastructural cooperation itself.

The cooperative movement, in taking hold in northern Newfoundland and Labrador, only gradually moved toward the self-granting of credit in 1940. Was that the "economic freedom" Grenfell was after for the fisherfolk? "It would not be the time to say that the whole co-operative venture has been an unqualified success," Grenfell writes in 1919, "but the causes of failure in each case have been perfectly obvious, and no fault of the system."[63] These "causes," Grenfell surmises, were created by two kinds of lack: of business ability and of "courage and unity which everywhere characterizes mankind."[64] The mission's credit was extended, and the fisherfolk of northern Newfoundland and Labrador took it up as best they could. "The debtor is 'free,'" Maurizio Lazzarato writes, "but his actions, his behavior, are confined to the limits defined by the debt he has entered into. The same is true as much for the individual as for a population or social group."[65] As such, the mission's infrastructural network of cooperative finance and its integration of credit granting became a merely reconfigured existential debt lived by the fishermen and their families—a form of debt inserted into a reformative "dynamic system of relations" and, so, the mission's own vital practices of infrastructural mediation. It was perhaps the best way toward achieving that ambiguous "fight for economic freedom" that Grenfell was after, if not one rooted in an equitable social exchange, both economic and symbolic. The debtor subjectivities of fisherfolk, disparate as they were, highlight how cooperation functioned as a form of life reliant on an uncertain Christian love, infrastructural care, and missionary credit. "Media not only bind (the past, the community) or network (space, time, people)," Peters writes, "but also organize."[66] Cooperation had to hold together a new debtor subjectivity across northern Newfoundland and Labrador that was organized to be, in time, self-eliminating. Their resource frontier held the possibility of gradually fading into a settler colonial reality of permanence, fading into an organization of life predicated on paved-over dispossession.

Spot-Cash Cooperative Company

As mission efforts to introduce cooperative action into various outport communities in the colony continued over the 1910s and 1920s, the Co-operative Wholesale Society (cws) of Manchester, to some extent the progenitor of

Grenfell's interest in cooperative economics, also pursued its own expansion. Working within a broad, inclusive, and class-based view of the working people of Britain, the CWS, by 1912, had a total invested capital of £2,590,218, supplies worth £7,556,821 at net factory prices, and a workforce of 13,370.[67] Despite achieving this scale, or perhaps because of it, Percy Redfern, undertaking to document the evolution of the CWS in a commissioned history, notes how this group of like-minded consumers and producers stood for an "industrial democracy—a direct service of working-class consumer by working-class producer."[68] He emphasizes how crucial it was for a network made up of artisans, clerks, laborers, and their mothers and wives to have "linked store to warehouse and warehouse to workshop and workshop to farm."[69] It was important not only for the commodity to follow a cooperative chain from producer to consumer but also for a group of working-class people, including women—an often-neglected domestic consumer in analyses of commercial distribution systems[70]—to enable a "voluntary collectivism" to take shape around the common economic practices of the British working class. Echoing Grenfell's struggles to inculcate a sense of unity within the fisherfolk of the colony, the CWS flourished because of a class-consciousness made operative in the face of the effects of the long Industrial Revolution.

One of the difficulties in bringing cooperative action to the colony rested in the nebulous cohesion of workers in the staple economy of the fishery. While the CWS worked across, to some degree, complimentary industries, the fisherfolk pursued livelihoods that, partly owing to skill, season, and luck, could vary widely from neighbor to neighbor. As such, fishermen were not easily gathered under the category of workers. According to Redfern,

> "All workers want four things," wrote Mr. Philip Snowden in the *Daily Mail Year Book* for 1913, "and if these four desires were gratified most of them would probably think that there was nothing more to be desired." These four were: (1) "assured employment," (2) "a decent wage," (3) "guarantee of provision in case of permanent inability," (4) "a reduction in the number of working hours." One would like to think that any not innately servile would desire at least one thing more: to feel himself a free and responsible agent, on terms of human relationship with his superiors, and not to stand to his task simply as a wheel of the machinery, a mere unit of business arithmetic.[71]

While the fishermen of the colony would most likely have welcomed structural conditions of labor such as these, they were, unlike industrial laborers, the ultimate "free and responsible agent[s]" who were reduced by the changeable circumstances of geographic remove, lack of transportation infrastructure, poor or nonexistent access to educational institutions, and the credit-capital heavy reliance of the fishery, to dealing on a one-on-one basis with local merchants.

Some labor arrangements, such as those deployed in the seasonal seal hunt (with hundreds of men working on a per-pelt basis on boats) and the merchant firm contract work taken up by some fishermen, collected groups of workers into wage-like relationships.[72] However, there were thousands of independent, family-based productive units engaged in fishing, drying, and salting (often done by women and children) on the coast, producing catches of cod to be sold to local merchants. As a result, the mission had to constantly strive to create a site-specific sense of voluntary collectivism that could convince the fisherfolk to take up the fight for economic freedom and cooperate, not in spite of their status as free agents, but rather because of it. With this goal in mind, the mission was often quick, through its newsletter, educational literature, and word of mouth, to point to its successes—its practices of infrastructural mediation had to be made to discursively cohere. By the 1930s the Spot-Cash Cooperative Company, based in St. Anthony, was doing an annual trade of more than $50,000.[73] St. Anthony, the Manchester of northern Newfoundland and Labrador, was also "the cheapest place in which to buy necessities anywhere in the North."[74]

Grenfell and the mission introduced several changes in the functioning of the day-to-day retail operation of the stores with the creation of the St. Anthony Spot-Cash Cooperative Company in 1913. The first of these was that nothing was to leave the store unless it was paid for in cash, a rule that was also to apply to members; in dire situations, members could receive necessities up to the value of their shares. The second was to appoint an outside manager, a man of "learning," and so a clerk from Toronto, working as a mission secretary at the time, was named the storekeeper and manager. The third was the creation of a stamp system devised so that members, in registering their dealings with the spot-cash store in common ledgers, would feel a sense of solidarity with other members. The stamps, with clasped hands as their device, were made by the American Bank Note Company of Ottawa.[75] The fourth, and possibly the most significant change, was that the store manager was not to give, under any circumstances, any credit to anyone. If he did so, he would incur it as a personal debt. Within the first two years of the store's existence, the majority of the changes would prove to be failures (the manager "'found it impossible to refuse the naked and the hungry'"[76] and extended nearly $10,000 in credit, with Grenfell, as a backer of the company, responsible to repay it to the creditors in St. John's). Yet, over time, and with the training of local young women and men sent to study at business colleges in the United States,[77] the spot-cash store became a retail fixture in St. Anthony.

The creditor-debtor relation was one that the mission not only sought to improve on but also one that displayed, in its evolution, the extent to which the IGA was after structural social reforms in the colony. At this macroscale of analysis, it is worthwhile to look at the lone appearance that Grenfell and the mission make in Innis's *The Cod Fisheries*:

> Improved transportation facilities by railway and steamship brought about changes in the internal market. The retail trade of St. John's tended to become less, as it was carried on by small merchants in the outports, but wholesale trade increased. Canadian banks which dominated Newfoundland banking after the crash of 1894 established branches in the outports and the cash system was gradually extended. "The people of the colony are becoming yearly more independent of credit from the merchant." These trends strengthened the basic decentralization of the economic and political structure of Newfoundland. The absence of local government except in St. John's meant that ecclesiastical organizations became active in education, and that medical and social services were extended by private agencies. Sir Wilfred Grenfell built up his hospital organization in northern Newfoundland and in the Canadian and Newfoundland Labrador. In 1896 he started a cooperative organization at Red Bay, which was followed by others at St. Anthony's, St. Modest, and Flower Cove.[78]

In Innis's reading, the IGA was a private agency managing the trickle-down effect of international finance, colonial and imperial markets of trade, and the dispersed nature of governmental authority and services in the colony. The IGA also managed what Innis deemed the "crude insurance system" that was the system of commercial organization under the truck system, which essentially sought to offset losses in one region with gains in another and, so, "losses suffered by some fishermen against the gains made by others."[79]

Innis was in part trying to soberly assess the effects, present and future, of a late-coming capitalism to the colony of Newfoundland and Labrador. Published in the same year as Grenfell's death, *The Cod Fisheries* was an appropriate marker for the changes that were to come to the North Atlantic fishery and the place of outport communities within it. "The fishing industry of the North Atlantic," Innis writes, "has been exogenous in its development."[80]

Small settlements gradually extended along the coastlines of the colony and maintained an outward looking relationship to the sea. This socioeconomic geography, for Innis, shows how "Newfoundland is a striking illustration of the effect of the impact of capitalism on a country's economy."[81] The mission's cooperative efforts show the parallel impact of this bourgeoning, and ultimately failing, entrepreneurial capitalism on the colony's outport fisherfolk. Innis, in the foreword to his book, states his commitment to political representation and the possibilities of an emergent system of commercial organization: "We cannot base our argument on the importance of the British Empire to the maintenance of democracy when we calmly allow the light to go out in Newfoundland. It is the hope of the author that this volume may contribute in its own way to a solution of the problems of the region by fostering that cooperation among its component parts which has emerged with industrialism, and by tempering the bitterness which has marred the history of the industry."[82] While Innis's sense of cooperation is imbued with the market-based fluidity of complementary sectors within a well-functioning industrial economy, it is nonetheless indicative of the sort of ideal evolution the creditor-debtor relationship could begin to take.

On the cusp of World War II, the mission's cooperative efforts in the 1930s were undertaken within a fishery that was to change decisively over the next two decades with the introduction and growth of the fresh-frozen industry.[83] The war years brought about a dramatic shift in wage-based labor in the colony as well as a cultural reorientation toward confederation with the Dominion of Canada and that state's governmental apparatus of social services and definition of public welfare. The early cash-based experiments of the mission, and their valorization of forms of voluntary collectivism, may have prepared the ground for the implementation of such a governmental financial technology as federal and provincial taxation. To work on the "collective" and to bring that sense of "unity" out of the realm of the debtor-debtor relationship needed, especially in Grenfell's evangelizing eyes, "love for the other fellow."[84] Religious affect was, as I addressed above, the emotive force that could triangulate between individual fisherfolk and the mission's cooperative infrastructure. Yet this love would inevitably come up against a series of obstacles across the spectrum of the staple economy, from the human failings of mistrust to the structural disadvantages of illiteracy and dependence on volatile international markets of trade. Grenfell's writings evince a duality between public hope in the cooperative movement and private realism, veering toward doubt. In a 1938 letter to Ashdown, he writes:

The poor result of the cooperative distribution effort at St. Anthony is due to the fact that all the neighbours were not able to forget their fears and venture to break from the old system and risk what that would involve, at any rate, at first. Fear is always the greatest factor in man's failure, and added to this the so-called struggle for personal advantages that is better expressed in the term "greed" than by natural wisdom cause many to yield to the irresistible argument, and yield to the temptation to seek immediate advantage rather than "endure" the penalties of sacrifice for a season.[85]

This makes for an odd resonance with the jubilant announcement, two years later, of the creation of the St. Anthony Co-operative Credit Society. While by this point Grenfell was a marginalized, more or less symbolic figurehead within the IGA, he was still one of the very few mission workers who could take a broad view of the cooperative movement's half-century evolution in northern Newfoundland and Labrador. This evolution, of course, also encompassed the more holistic situation of the staple economy, and so the coming of cash to the colony in some fundamental sense also started to erode, in Grenfell's projected social future, the coming of the values of the Christocentric social gospel. If, as Gilles Deleuze has remarked, "beyond the state it is money that rules, money that communicates,"[86] it could follow that the near simultaneous implementation of a cash economy, an evolving industrial capitalism, and state-sanctioned aid and control in outport communities in northern Newfoundland and Labrador all benefited from, and possibly co-opted, the mission's preparatory work in commercial organization. Infrastructural mediation, in this instance, did build out material forms of life that worked their way toward a capitalist end after World War II. The mission's credit came with existential stakes after all.

In the summer of 1934, after several years of failed attempts, Grenfell finally managed to bring Margaret Digby of the Horace Plunkett Foundation to outport communities in the colony.[87] Digby, a major figure in the British cooperative movement and a promoter of cooperative organization through active research, published her findings in *Report on the Opportunities for Cooperative Organisation in Newfoundland and Labrador* in the same year as her visit. Digby's trip was sponsored by the IGA, the Commission of Government of Newfoundland, and the Wholesale Co-operative Society of Manchester. At the time, Grenfell interpreted the Commission of Government's support as "endorsing the decision to which we came many years ago, that the producer of wealth has a right to make it go as far as he can, and to buy as well as sell in the best markets."[88] There were varying interpretations of Digby's report. In the view of the Commission of Government, Digby laid the foundations for the creation of a cooperative division in the Department of Natural Resources in 1937 as well as establishment of a co-op act in the colony of Newfoundland.[89] Grenfell, in a 1937 letter to the editor of the *Observer's Weekly* of St. John's, states:

> At the risk of making this letter too long, this brings me to another point, the word "cooperation." Perhaps you are acquainted with the report of a specialist in cooperative efforts, both productive and distributive. Some years ago we sent Miss Digby together with the cooperative leaders in England out to report on the possibilities of the effort at Red Bay where all the people joined in the cooperative effort to send all their fish up together and bring back the value in supplies is, I still believe, functioning and successful as far as the effort, but the report was that Newfoundland was not yet ready for cooperation, though in England as you must well know, the tremendous value and success of that movement.[90]

In the span of three years a realpolitik with regard to cooperative possibilities has Grenfell, both publicly and privately, looking to a more uncertain economic future. Without substantive protective legislation and general cooperative education, as Grenfell claims in a 1934 letter to R. A. Palmer of Manchester, a member of the Manchester Co-operative Union and vice president of the International Co-operative Alliance,[91] cooperation would remain a nascent practice of fisherfolk life—poorly managed and without maintaining a sense of infrastructural care as well as a religious affect of unity.

On the whole, throughout the 1930s, Grenfell and the IGA worked on extending a form of credit of their own devising. "Granting credit requires one to estimate that which is inestimable—future behavior and events—and to expose oneself," as Lazzarato writes, "to the uncertainty of time."[92] It was at multiple scales of the staple economy, as Innis diagnosed, that this mission credit was renegotiated, recast, and evolved into a form of post–World War II industrial capitalism that Grenfell would have been at pains to recognize. Such practices of infrastructural mediation are indeed tied to the shifting conditions of time. What cooperation shows us is that these practices of mediation contain a politics of duration that are tied to their existential stakes: what forms of life they enable, constrain, or do away with and what infrastructural commitments they make to change the everyday material circumstances we inhabit. The promise of extraction that kept the settler colonial fisherfolk in

Sir Wilfred Grenfell,
Kinloch House,
Charlotte, Vt. U.S.A.

Red Bay, Mar. 26, 1938.

Dear Sir Wilfred:

I received your letter dated January 19, 1938, and was moved to tears on receipt of the present enclosed, for which I cannot find words to express my thanks to Lady Grenfell and your good self. I will treasure the memory of that gift, while I live by using it to help the unfortunate poor, as I know you would do if you were here and my prayer is that God may bless you both.

I will try and answer the questions you asked to the best of my ability. First. It will be six years ago the coming summer since I saw you.

The Co-operative Store has been operating since Aug. 8, 1896--that will be forty-two years August 8, 1938.

I hope you do not give me all the credit. I have only done my duty as far as I knew how, with some mistakes I'll admit. I have tried to forget self and follow the teaching of the Master, inasmuch as you did it to the least of these my Brethren, you did it to me.

As I look back over the years to 1896, the only thing that grieves is that I did not do better. There are only four of the original shore holders left, including yourself. These are my father, Ephraim Yetman and myself. You asked me what I thought of the Fishery. As far as I can see, its rapidly declining and has been since the tidal wave of 1929. The Fishery has been a total failure on both sides of Straits from Flowers Cove to Bonne Bay on the N.F.C.A. side and from Bradore up on the Labrador side.

The Cod Fishery has been very poor around here and Forteau, but better here than last year. This is partly due to economic conditions, people unable to replace their fishing gear and dog fish debt in August of this year hindered us from procuring an average voyage.

(Question 1 on Second sheet) I cannot answer this question. I did not know myself that the other stores started by you were on a different basis.

(2) I will try and do my best to talk co-operation with any one interested and give all the information I can in regards to methods used. And to my mind it is very easy to get the key. It is this, treat every one right - buy all their products and treat it ascash, give them St. John's price for all products less freight and insurance. Keep overhead expenses as low as possible. Sell goods, provisions as cheap as is possible to the consumer and give every man an detailed account at the end of the year, so that he can see for himself where his earnings go.

(3) I hardly think Sectarianem has anything to do with

FIGURES 2.2 AND 2.3. Letter from William Y. Pike to Wilfred Grenfell, March 26, 1938, MG 63.164, Grenfell Letters, International Grenfell Association Fonds, Grenfell Association of Great Britain and Ireland, Provincial Archives of Newfoundland and Labrador.

co-operation. The Red Bay Store has been supplying ten families in Lance liu Coup for four years now with salt, gas and provisions, also several families - Lance au Drable, West St. modeou and ponvase. I have not had any trouble - ninety five percent are honest and straight, certainly there are few dishonest, it must be so. Treat men as men and brothers and they will certainly reciprocate.

The Co-operative Store has been in operation for forty-one years and nine months and in 1937 its turnover was approximately $18,000. It paid this year $4.50 per quintal for fish when the current price in straits was $4.00 per lb. So you see Red Bay store paid on what fish we bought 2300 quintals $1150.00 more than the traders did. We paid $2.00 more a cask for Cod Oil, about 20 percent more for furs they bought, summing up the year's work, The Co-operative Co. owe no man anything today. I am now sending an order for supplies and salt for 1938. We have in stock now, wholewheat flour, Cocoa, tea, sugar, corn meal, pork, beef, oleo, prunes, etc.

Seeds that are hardest to get is cabbage and Turnip (Swede) People of this community took 200 barrels of potatoes from the ground last year, these were a good help to people if they had to buy them out- side it would cost us $500.00. We have among us 50 sheep and about fifteen head of cattle and quite a number of hens.

My father is alive but getting very feeble now. He claims he will be 99 Sept. 1, 1938. Geo. Ash is well, age 80. Richard Cannory well, age 86. Moores families are all well. Minnie is well and was with us today to dinner and tea.

I am striving to forget self and carry on to the best of my ability so that when I pass on to leave a record to the people of Red Bay and to my successor of straight honest dealings, and the Co-operative Store in good standing as a memento to your kindness and generosity in making it possible for us to do our own business and save middle-man-profits.

I received post card from you on November 1937 but was per- suaded not to write you on account of your health. Our prayer is that you and Lady Grenfell will be spared to see many more years to come.

Yours very sincerely,

William Y. Pike

P.S. Please write again and I will answer to the best of my ability.

the colony was not so much mediated into being as it was a promised form of life held together by the prospect of debt, a settler practice immersed in the ambient cultural, financial, and social—in sum, infrastructural—environment it inhabited. A resource frontier and environmental medium on its way to becoming settled infrastructure—from chalk to paper and wood.

The mission's practices of cooperative reform constituted a vital infrastructural mediation predicated on expectations of mutual aid and the granting of credit as well as a tangible infrastructure of care made up ledgers, ships, and stores. It is worthwhile to end by returning to the photograph of the Red Bay cooperators of 1896. Over the course of the mission story, no other cooperative store had such an even-keeled economic journey as the Red Bay Cooperative Store. Weathering poor fishing seasons, price fluctuations in international markets for goods both bought and sold, a shift from European to Caribbean and South American buyers, and service in World War I, among a host of other structuring influences, Red Bay gradually negotiated a form of existential debt that could equally evolve with their proximate circumstances and the global economy. Amid Grenfell's moral pessimism, public posturing, and evangelical turn in the late 1930s around the question of cooperative action and its legacy in northern Newfoundland and Labrador, on January 19, 1938, he wrote a questionnaire of sorts to William Y. Pike, the original, long-serving manager of the Red Bay Cooperative Store. In response, Pike, one of four original and still-extant shareholders, takes Grenfell's questions in turn, relating the precise opening date of the store, August 8, 1896, nearly forty-two years in the past, the poor fisheries of the past several seasons, and going on to provide a detailed summary of its commercial operations: paying $4.50 for a quintal of fish while the price in the Straits was $4; paying $2 more for a cask of cod oil; and on 2,300 quintals of fish, paying $1,150 more than the traders did. "We have in stock now," Pike writes, "wholewheat flour, cocoa, tea, sugar, corn meal, pork, beef, oleo, prunes, etc."[93] With Grenfell asking Pike to reflect on the years that have passed (as we can only surmise, given that Grenfell's original letter does not survive), Pike is careful to recall the creditor-debtor relationship, both moral and economic, that underpinned the store's longevity and its outlook: "I hope you do not give me all the credit. I have only done my duty as far as I knew how, with some mistakes I'll admit. I have tried to forget self and follow the teaching of the Master, inasmuch as you did it to the least of these my Brethren, you did it to me."[94]

Agnes Patey: As a child I had TB spine. From the time I was two until I was thirteen, a lot of that time I spent in the hospital. I really benefited from the care Dr. Grenfell and the people who came after him brought to this area. . . . I was born in '46. When I was 2, by that time Dr. [Gordon] Thomas was here. . . . My mother did work at the hospital for a short time before she was married. . . . I've been involved with the historical society since 1985. He [Dougal Dunbar, designer of the Grenfell Interpretation Centre exhibition narrative] gathered all the information, and put together the whole design of it. Then he contracted it to another company to do the actual storyboards. Most of the material was already here with the historical society. There was a committee in place. And we identified the material. He laid out the story. The sequence of the way things went. We tried to interpret what Grenfell brought to the area.

Frances Patey: He [Grenfell] liked to let people know that he was calling the shots. I did that book there, *The Grenfell Dock*. That's the diary of the Grenfell dock. When he was still active over there at the dock, everything connected with that dock and what they did was written. After he went I took notice of how that started to deteriorate, there wasn't so much written up. The book is an actual log of the Grenfell dock. . . . Every schooner that used to go up to the Labrador fishery went up on that dock for repairs. Over two thousand when the dock was there.

A: Grenfell wouldn't have had much to do with the dock by the time that was done (he was long gone).

F: What's the day marked that it opened, marked there on the cover? 1928.

Models in Francis Patey's basement workshop. St. Anthony, Newfoundland and Labrador, 2011. Photograph by author.

A: But the last year he actually lived here was 1920. He died in 1940.

F: I believe the last time he was here was 1938.

A: Yeah, that was his last trip here.

F: That was his last trip here. . . . I have six brothers, my father, mother, every one of us, even if it was just for one month, worked for the Grenfell Association. My mother even served him when she was a girl, I think. When she was very young. As a servant girl, eh. And one of my brothers, he died recently, he was 56, when he was a young boy he had something wrong with his eyes. There used to be a tourist boat would come here, called the *Northstar*, it had a hospital on board so Grenfell took my brother there and got his eyes done. Then he wanted him to go down to the States to train for something, and he did that with a lot of boys around here, not a lot but some. But my father wouldn't approve of it so he didn't go.

[. . .]

A: For me it's hard to tell. Everyone benefited from the mission, but there was always kinda that feeling that the people who worked for the mission considered themselves a little better than everybody else, and that kinda thing was there in the community. In the early days, everybody certainly benefited from it and realized how lucky they were, at least to have a job and everything else, but there was always that bit of . . . because people came from away, so many people who worked there came from away, and there was a little bit of a divide between the actual mission people and . . . 'cause people who worked there were called "mission people" . . . I guess there was probably a bit of jealousy there, which would be human nature kinda thing.

F: Back then, it wasn't called anything else; if you mentioned "mission," people knew what you were talkin' about. "Where you work?" "Over on the mission." Wasn't called with Grenfell or nothing. Was just called "the mission."

A: Well, that's what it was at that time.

———————

F: The same thing. I spent some time up there in Nain, Makovik, five or six years. I worked with social services in Nain. I taught school in Hopedale for two or three years. Back in the '60s. And Grenfell was there with the nursing stations. Nain was the biggest one then.

———————

A: A lot of things that were associated with Grenfell . . . Grenfell started it, he was the impetus to do it, but I mean other people picked them up and continued on, and did that sort of work. There was a lot of that kinda thing. He had all those ideas that he formulated, but other people continued the work. I've heard people say that he was really scatterbrained, kinda like. He was going off in all directions to get things done. But of course he actually depended on other people to carry it out. . . . I often think, well, how could one person accomplish so much in a lifetime, with all the things that he did and associated with him doing it, but for the most part, for so many of them, they were just his ideas that he got started, and other people carried them on.

———————

F: Young people from Labrador, he brought them to the States to be trained. His number one man, his architect, guy by the name of Teddy McNeil. He built whatever you see over there. Not the new hospital, but the orphanage and that. That was Teddy McNeil. He brought 'em there, and trained him in the States. And he did all that. Them coal bins, that was Teddy McNeil.

A: But a lot of the things he did do are not around anymore. A lot of it has been overbuilt, torn down.

F: That's a big shame. They destroyed so much good stuff . . . history. The dock, the Grenfell dock should still be sitting over there.

A: But Francis, that's like with everything else; there was no one to maintain it. You can't keep something here just because of what it is.

F: Down in St. John's they keep something if it's only two years old! The Murray Premises . . . have you been to the Murray Premises? It's all from the St. John's area. I didn't see anything from down this way, or war veterans or nothin.' There's nothin.' The stuff they got there doesn't really spread out over the whole province.

———————

A: Grenfell himself, he would go out into the communities, and then as the years progressed, the public health nurses would go into the communities and do public health work. They'd go by dog team. Grenfell used to make his winter trips, to Flower's Cove, for example. In the summer, there were the hospital boats.

———————

A: When I think about home, as a child, I think about the hospital. I don't think about my parents' home so much, because I spent Christmas and whatever in the hospital. To me, the Grenfell story is something that really needs to be preserved. Because right now, in the health care system, it's not Grenfell anymore. Grenfell's attitude was that he took care of everybody. But the attitude these days is the government attitude. There's so much management involved with it. It's not a hospital anymore; it's a managed building. His attitude was a holistic attitude towards the whole individual. I know times change and everything is different now than in his time, but his attitude was that people had to have work; they had to take care of their bodies and their souls, everything to try to help the whole individual. Of course now it's just totally focused on health care. For me, because I was raised in the hospital, I see so much of the difference. Someone else wouldn't even notice it. I was not in Grenfell's time; he was dead before I was born, but the Grenfell attitude was still there in the hospital. [In the 1950s] they still had cows and pigs that supplied the hospital, not so much gardens at that time. I can remember when they decided the cows had to go, because obviously it was in the middle of the community. They brought in what was called a "mechanical cow," which made powdered milk instead of having fresh milk. That was in the late '50s, early '60s.

F: The reindeer came first. Came over from Lapland. The government every day is trying to draw away a bit from Grenfell as we knew it.

[. . .]

A: There were so many people who came from so many different countries. There were teachers and doctors, every profession that you could imagine. Their cultures, especially in the schools, would have had an effect on the younger people. They had much better quality teachers than probably a lot of other smaller communities would have had.

Regarding the proposed tunnel under the Strait of Belle Isle to connect the Labrador Highway to the island of Newfoundland.

F: There's a little Newfoundland joke; we got one, ya know. Wanna hear it? One fellow up the Strait says what about they're going to start on both sides, Labrador side and this side; one old guy said, "What about if they don't meet?" He says, "We get two tunnels from the price of one."

F: Another thing. Grenfell started the first co-op in Newfoundland, I think.

A: Back in Grenfell's day everything was wireless telegrams. I remember this one little girl—in fact she's still living here in the seniors home—she was the messenger girl. She would bring the messages up from the building to the hospital on her bike. I can remember that. That would have been in the '50s. The messages would come in, and she would bring them up. Of course the ones that had to be sent out would be the same thing. She'd bring them down. That was basically the only communication at that time. Everybody wrote letters and cards too.

[. . .]

My mother had a sister; she was a nun, and she lived in Nova Scotia, and she was only allowed to write letters home one Sunday a month.

F: When the mail planes started, in the '50s, there was one every two weeks. I wrote a book on the bush pilots. Some of them are still living in the St. Anthony area. No aids to navigation or nothin.' When they couldn't see their way into the community, they smelled their way in. There was no communication.

———————————

A: The thing about the Grenfell Mission is that whatever was the latest in technology, they had it. As things became available, the supporters would supply it to the mission. Even radiation treatments; I think the first ones were done here.

F: They say Dr. Thomas was the first person in Newfoundland to take the skull off a person in an operation.

A: I don't know if that's true or not.

F: That's what they say.

A: Thomas did heart research on dogs here too. I don't know if he ever perfected a procedure, but he did it. He had a little house over there, and he used to do the surgery on dog's hearts.

F: He was a very, I don't know, stern man. He was a man who never got too mixed up with the community. He was a man you almost had to get an appointment with to say hello [laughter]. He was a nice man obviously. But he was strictly a doctor. I guess the biggest story of the Grenfell era is when he almost messed up on that dog team that time.

The families of those who picked him up still live here. Grandchildren, great-grandchildren, and that.

————————

A: I'm someone who's interested in keeping the story alive.

F: CBC [Canadian Broadcasting Corporation] called me once; they addressed me as an "author." And I said, "I'm not an author." They said, "You writes books, you must be an author." No, I said. "I'm somebody who marks stuff down so I won't forget it."

————————

A: The Grenfell Interpretation Centre got put here at the time of the cod moratorium. So many people lost their jobs of course, so the government gave money to put people back to work. They had to recognize that the Grenfell story and the history was worth preserving. But that was the main impetus to get the funding to do that, to get the actual building. We were told, at the time, that we had to find ways to maintain it. They wouldn't be running it or anything. The store brings in a fair bit of money, the tourists in the summer time; of course we don't have a lot of staff, just keep it to the bare minimum. We really need a historian, a professional, but we just don't have money. To be honest, I don't know if we'd ever get anybody to come here because the season is so short, with July and August the two good ones. What would the professional be doing in the winter months? There's not a lot of work to keep you busy.

————————

F: Back when I was growing up, all the mothers would hook mats for the Grenfell. And they used to have a clothing store over there, with clothes coming in from England and the United States; people used to donate, so my mother would take the mat over there, and they would give her clothing for us for the mat. There are still hooked mats now.

A: A Grenfell mat is identified by the fact that every hole is hooked. Mat hooking died out for a lot of years because it was always associated with poverty—like Francis just said, his mother would hook a mat, and you wouldn't get money for it; you'd get clothes. When people became a bit more affluent, of course, they didn't want to be doing that; that was associated with the poverty. So for a long time there

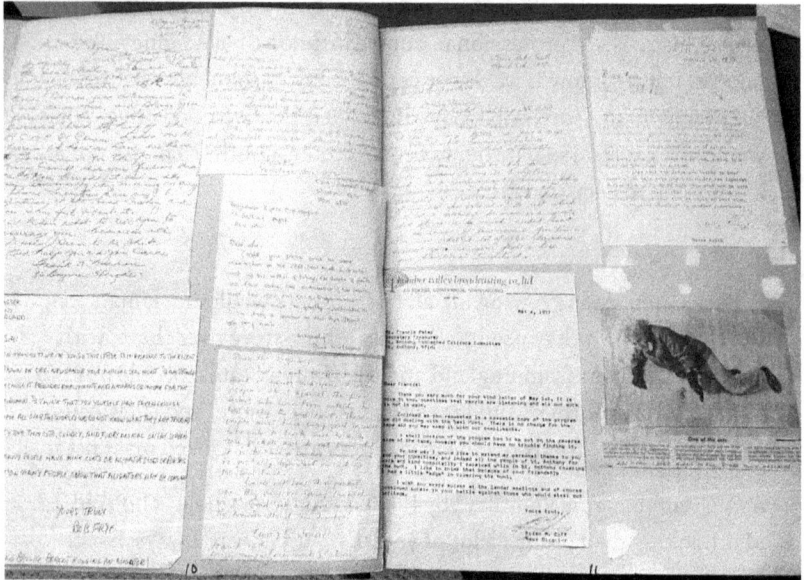

Francis Patey's scrapbook documenting his efforts to save Newfoundland and Labrador's seal hunt as a member of the St. Anthony Citizens' Committee in the 1970s. St. Anthony, Newfoundland and Labrador, 2011. Photograph by author.

was no mat hooking. In recent years it's come back. People have taken it up as a hobby. The historical value is recognized. St. Anthony has been, for the most part, a prosperous community over the years. It's been a community with a fairly well-educated people.

F: The Americans moved in there in the 1950s. Over two hundred. They were here for fifteen years till their base closed down. Nothing left but the concrete. They cleaned it all up. All the sites throughout Newfoundland and Labrador. The Pine Tree Line is all cleaned up now. I think there are a couple up in Hopedale that aren't completed. There were five or six in Newfoundland and five or six in Labrador. Every so many hundred miles. There were large ones and smaller ones in between. Most everybody here who didn't work with the Grenfell Mission worked there. Everyone here called it "the site." "Everybody works on the site." Some of the soldiers still come back.

3. META INCOGNITA

In a note dated February 27, 1931, Alexander Forbes, a professor of physiology at the Harvard Medical School, recounts a recent visit by Ellery F. Atwood, a carpenter by trade, of 240 Warren Street, Allston, Massachusetts. Atwood informs Forbes that over the past several months he had begun to "receive messages and visions imparted to him by spirits."[1] As he tells it, three weeks earlier he was sitting with his wife in the early evening when a vision appeared of an airplane spinning out of control. Coming down over the ice was the plane of Charles Nungesser and François Coli, French aviators who had attempted to cross the Atlantic in 1927 (preceding Charles Lindbergh's crossing by two weeks) and whose plane had essentially disappeared without trace. "It was approaching the Atlantic coast," as Forbes relates, "and went into a dive and came down on a narrow strip of land called Meta Incognita." Assuring Forbes that he had no previous knowledge of the geography of that part of the North (Forbes: "I asked him if he had studied the maps of that region prior to his vision"), Atwood goes on to draw a diagram of the precise location of the plane, north of Hudson Strait, south of Baffin Land; "the wreck of the plane would be found on Meta Incognita between the Great Crevice and Resolution Island."[2]

Alexander Forbes was a busy man in February 1931. In four short months his first surveying expedition to northern Labrador was scheduled to begin. Over the course of the preceding three years, the idea of mapping the northernmost coast of Labrador taking shape on a suggestion made by Grenfell to Forbes in 1928, he had bought and outfitted an auxiliary schooner, the *Ramah*, as well as a Fairchild seaplane, specifically designed for aerial surveying and was in the midst of planning the minutiae of the expedition: rations, surveying equipment, crew responsibilities, and numbers. Moreover, he had to take into account the broader aims of the geological and botanical work that would supplement that of the aerial photographic survey sponsored by the American Geographical Society (AGS), based in New York. As such, Atwood's visit and spiritualized mapmaking was an exercise in the absurd for Forbes. This absurdity stemmed from the fact that the entire premise of the Forbes-Grenfell Survey of Northernmost Labrador, the official name adopted by all concerned parties, was to supplement the minimal existing maps and charts of this part of the coast. The expedition was after a context-dependent understanding of *accuracy*, a mobile variable made up of cost, time investment, and the real value of the territory in question. Through the use of an experimental surveying technique relying on oblique aerial photography to map regions of high relief developed by O. M. Miller, a British-born surveyor with the AGS, the expedition was seeking to map what was then a still poorly known (to settler eyes) section of the Labrador coastline as well as parts of its interior. Yet Atwood's vision, read in a certain charitable hermeneutic light, speaks to the porous boundary between the then-prevalent pragmatic and leisurely uses of airplanes; the relationship between remote, previously inaccessible regions and the new geographies made accountable, through mapmaking, if not actually accessible by the medium of the air; the place of flight in the popular American imagination, one occupying a combination of adventure, danger, speed, accomplishment, and resolve; and, finally, the misty, inaccurate knowledge surrounding such sites as Meta Incognita, the "unknown shore" in both the minds of a carpenter such as Atwood and a physiologist and amateur aviator-explorer such as Forbes as well as in the tangible world of mapmaking. Atwood, wanting to help locate the previously unlocatable evidence of a trans-Atlantic flight, was also pointing to what could lie within the purview of those willing to go to such unknown shores to make a map.

Forbes: "He said that until two and a half months ago he was an agnostic and had no belief in spirits, but the thing had come to him suddenly and he was now a changed man and his feelings were all different. He doesn't expect others to believe it, but he says he himself knows, and he has no motive other than the fact that the things are given to him to communicate."[3]

THE DEFINITIVE NARRATIVE OF the survey, reified in the book *Northernmost Labrador Mapped from Air*, published in 1938 by the AGS, gives the official account of what went into mapping the region.[4] It is a remarkable artifact, with its most striking documents being the map sheets included in a small pouch glued to the inside front cover. The diagrammatic charts, in addition to representing detailed and contoured coastlines, also trace the expedition's flight lines. They track up and down the mountainous coast, with occasional small circles and directional arrows indicating where and in what direction a high oblique photograph was taken. These tracings belie an infrastructural relationship to northernmost Labrador and one that was documented in the moments of its making.

While Forbes would write to Prime Minister Squires that "I am most anxious to have the expedition be as useful as possible to the people of Newfoundland and Labrador,"[5] including the fisherfolk dependent on the Grenfell Mission, their efforts toward creating a "best map" had among its ultimate and potentially most long-lasting effects the perfecting of oblique aerial photography as a surveying technique that would, in time, be integrated into military planning, particularly the U.S. nuclear-testing program during World War II. This is a consequential overturning of well-established narratives of technical "advance" that often stem from military use and trickle down into civilian life; Rey Chow writes, "As a condition that is no longer separable from civilian life, war is thoroughly absorbed into the fabric of our daily communications— our information channels, our entertainment media, our machinery for speech and expression."[6]

The mission's deployment of oblique aerial photography through Grenfell's provision of the initial impetus and logistical support to Forbes's expeditions, was part of its practices of infrastructural mediation and one that would be placed at the heart of American military technical life from 1939 onward. Aerial photography could come full circle, back to its beginnings as aerial reconnaissance in World War I. As a Fairchild Aerial Camera Corp. pamphlet of the time states, "Like the airplane, whose value as an instrument of war led to its early development, aerial photography has been fostered chiefly because of its military usefulness."[7] The same Fairchild Model F-4 camera that the surveying expedition would put to work in Labrador was also "used exclusively by the United States Army and Navy, the Department of National Defense of Canada, the Imperial Japanese Army, the Air Ministry of Brazil, Peru, and other progressive governments."[8] However, the settler-civilian and missionary narrative surrounding oblique aerial photography's use in the mapping of the North Coast of Labrador is distinct in its being the primary testing ground for a technical medium that would become essential to militaristic pursuits.

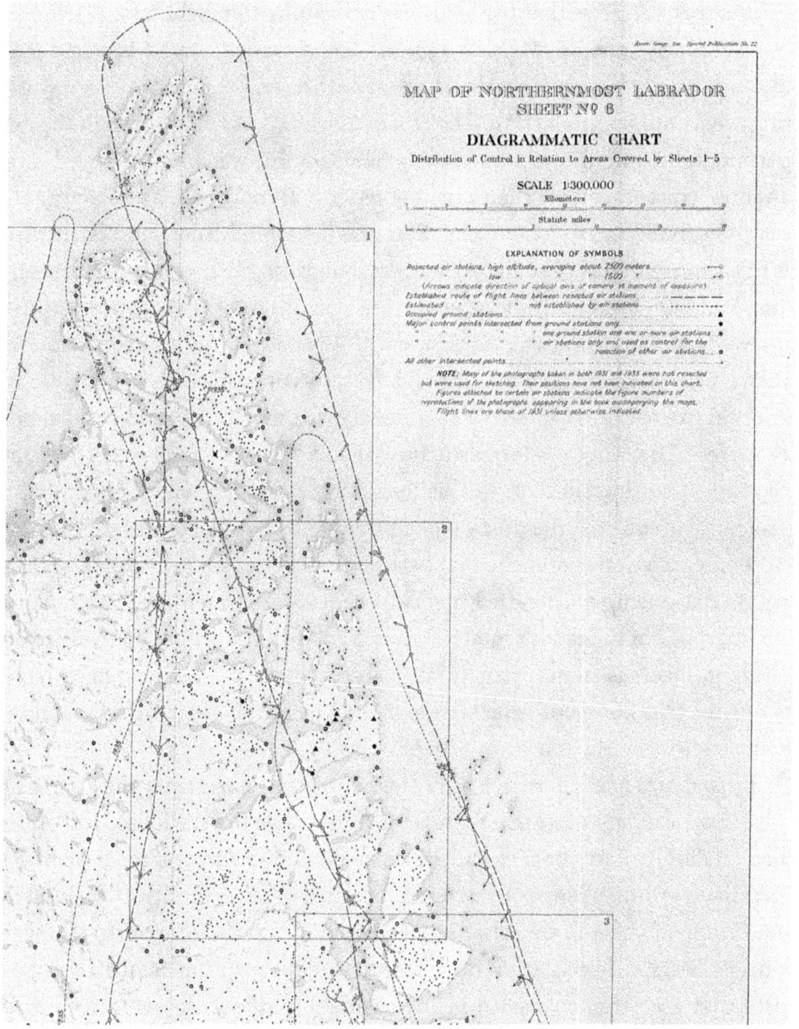

FIGURE 3.1. "Map of Northernmost Labrador." Air Sheet no. 6, in Forbes et al., *Northernmost Labrador Mapped from the Air* (New York: American Geographical Society, 1938).

It is this narrative that I will put under examination here. The 1938 book *Northernmost Labrador Mapped from the Air*, remarkable as it is as a documentary medium, also conceals, as remarked above, the infrastructural work that went into its production. The North Coast of Labrador, through the experimental medium of oblique aerial photography, was being mapped into an infrastructural rather than a merely territorial condition. As Ashley Carse observes, building on Susan Leigh Star and Karen Ruhleder, "Infrastructure is not a specific class of artifact or system, but an ongoing process of relationship building. Seen in this way, engineered canals and highways are surprisingly social and ecological. As temporary lines across active environments that erode, rust, and fracture them, infrastructures advance and retreat in relation to the capital and labor channeled into their construction and maintenance."[9] The mission's infrastructural mediation was predicated on a conception of settler life that, in this instance, was grounded in the evolving ecological and territorial unknowns presented by the North Coast of Labrador—an active environment that was open to being set into an infrastructural relationship with varied media of capture. Between the high oblique photographs of Labrador's mountainous terrain and the flight lines traced on the map sheets is an infrastructural, processual, and relationship-dependent narrative that flows out from settler-civilian technical life toward the militarized ends of World War II.

This chapter unbinds not only the competing narratives that underlie *Northernmost Labrador Mapped from the Air* but also the static medium of the book itself. It charts how the mission's use of infrastructural mediation was indeed environment-responsive and proceeded in tandem with O. M. Miller's reworking of the economies of aerial surveying and the American military's nuclear tests at Bikini Atoll. It is essential to open up the competing interests that are often concealed in the authorship of such a definitively true topographical statement as a map.[10] Infrastructural mediation, particularly as it was understood by the mission, is a practice that can point to the ways in which, as Parks and Starosielski contend, "infrastructures and environments dynamically mediate and remediate one another."[11] In this case, practices of mapping that were facilitated by the mission served to render the North Coast of Labrador into an infrastructural relation that was contingent on tensions between its own ecological vitality as a *meta incognita* and its co-optation into a technical instrumentality that was part of oblique aerial photography's narrative of civilian-to-military use.

FIGURES 3.2 AND 3.3. Oblique aerial photographs of the North Coast of Labrador included as part of O. M. Miller's ground survey report. Box 100, folder 2187, Alexander Forbes Papers, Francis A. Countway Library of Medicine, Center for the History of Medicine, Harvard Medical Library.

Planetabling from the Air

After years of preparations, on June 16, 1931, the *Ramah* left Lawley's Yard in Neponset, Massachusetts, laying its course for Brazil Rock off Nova Scotia. Reaching St. Anthony, Newfoundland, on June 29, the *Ramah* continued up the coast of Labrador, reaching, on July 18, a then-unnamed fjord (eventually named Seaplane Cove) that the party decided to make their headquarters, as it presented favorable conditions for the takeoff and landing of the seaplanes. As Forbes recounts, closing in on the fjord was a hazardous exercise because they were beyond the scope of any reliable maps. Relying on Grenfell's sketch map, a "little penciled sketch" as Forbes put it in *Northernmost Labrador Mapped from the Air*,[12] they marked their arrival at their chosen base by assembling the entire crew for a meeting to once again go over the objectives of the expedition.

The British geologist N. E. Odell gave a short outline of the geological features of the mountains in Labrador, and Ernst Abbe, a Harvard botanist, explained what he hoped to achieve. The majority of the expedition's members, quite nearly all amateurs, were devoted to ensuring the successful outcome of the surveying work. Forbes's position was that of skipper. The mate was F. Trevor Hogg, an architect and experienced sailor who had helped Forbes plan the trip from 1928 onward. The radio operator, also the youngest member of the expedition, was seventeen-year-old Edwin D. Brooks Jr. The ship's surgeon was Dr. Harrison E. Kennard, a recent graduate of the Harvard Medical School. O. M. Miller's assistant, James L. Madden, was both an experienced sailor and an amateur surveyor, having taken one course in the subject. The engineer, Donald J. Smith, had served on numerous Grenfell Mission boats. The cook, Charles Hatcher, a Newfoundlander, had been a cook on Gloucester fishing schooners for a number of years and so was a fair sailor in his own right. The rest of the amateur crew consisted of Brewster Morris, Wilfred T. Grenfell Jr., and Robert S. Hurlbut. Also, Hogg's wife, Odell's wife (both largely unnamed in the archival record), and Forbes's teenage daughter Katharine came along as part of the expedition.[13]

This is the coherent and teleological expedition story that Forbes tells in his 1938 monograph. Yet, in a 1936 draft of his opening paragraph of *Northernmost Labrador Mapped from the Air*, Forbes reveals what was a prevalent if often poorly acknowledged cultural assumption when it came to places such as Labrador:

> The purpose of this book is to tell the story of a mapping project at the northern end of the Labrador peninsula. The question has been raised,

"Why should one go so far afield to map a desolate, uninhabited and (to most people) useless region?" This question is hard to answer logically. Actually the answer was found in the haunting lure of the project, with its promise of scenic grandeur, and in the number of persons who showed eagerness to go on the expedition when they heard of it . . . , [it] might mean the saving of lives and property.[14]

In this version Forbes largely sees Labrador as a cruising destination. Moreover, Forbes would first publicize his expeditions to Labrador in such popular publications as *Yachting*, and so, in this early draft of what the AGS saw as a scientific publication, he was writing for a projected public of amateur "cruisers" potentially eyeing Labrador as their next summer's destination. The AGS's editor, Raye P. Platt, acknowledges "that having written a narrative once it is extremely difficult to rewrite it from a different angle."[15] Nonetheless Platt would ask Forbes to reorient his narrative of the expedition "from the point of view of the contributions to geographical knowledge" that it had made.[16] Labrador, far from being a "useless" territory, was an experimental site of geographical knowledge production.

As published, the book's opening paragraph reads,

> This book tells the story of how a map was made of the northern end of the Labrador peninsula. The question has been raised: why go so far afield to map a desolate, uninhabited, and, to most people, useless land? There are several answers. The bold and rugged landscape, the intricate coast line, and the lack of forests made northern Labrador an ideal region in which to apply and perfect certain new methods of aerial mapping. . . . The charts and sailing directions are inaccurate, based, for the most part, not on surveys but on such miscellaneous information as the reports and sketch maps of earlier voyages. . . . If these were not reasons enough, another—less logical but perhaps even more impelling— was the haunting lure of a wild and almost unknown country.[17]

In this reorientation of Forbes's original framing of the expedition, Labrador becomes an "ideal region" for scientific advance. The impetus provided by the unreliable state of piecemeal mapping projects has supplanted the lure of the unknown. Rather than merely approach Labrador as a territory to be mapped, the expedition placed the region into an infrastructural relationship with the experimental technique of oblique aerial photography. The North Coast of Labrador was to serve as a geographic placeholder (as outlined below) for the financial (lower cost), geographic (larger coverage of mappable territory), and temporal (quicker rates of capture and production) forms of compression the technique would enable.

In the preface to his book, Forbes lays out the genesis of the project in a more straightforward manner: "Originally planned as a summer vacation cruise to be enriched by a minor geographical objective, the project with which this book deals developed into a serious research in new methods of surveying from the air and absorbed a substantial part of the time of a number of workers for seven years. The geographical results include, besides the maps published herewith, material for the mapping of a large additional area."[18] This view of the expedition as one of happenstance and momentum is in fact borne out by Forbes's correspondence over the period running from 1928 to the early summer of 1931. The priorities of the expedition seem to emerge through social networks, common readership of the *American Geographical Review*, the availability of botanists and geologists, and the funding that would trickle in from interested parties, whether geographical societies or individuals with an interest in prospecting for minerals in Labrador. In addition, as Forbes alludes to in his final phrase, given that mapping as a practice is inherently open-ended, he hints at the realization that there will always and inevitably come a time when a better map will need to be made. Such a projected future of increased accuracy stems, in part, from the more or less equally open-ended, geopolitical futures that such remote and so-called useless sites as the North Coast of Labrador engender in the imagination of a variety of actors (both states and individuals); this in turn was borne out during World War II and into the Cold War when Labrador, and Newfoundland as a whole, would take on a strategic interest as the North American continent's first line of defense.

What also seems to get at least partly lost in Forbes's narrative is the value the surveying work could hold for the mission. While the latter was not the principal backer of any of the expeditions—a position held by the AGS, which would retain copyright over all materials resulting from the three expeditions—the mission nonetheless did have a vested interest in the work. After all the back and forth between AGS editors and Forbes, the book's opening paragraph proceeds with the following declaration: "The project really had its inception in a conversation with Sir Wilfred Grenfell some years before the expedition set out. Our talk was of cruising and sailboats. 'If you like cruising,' he said, 'why don't you come and map one of the uncharted fiords in Labrador?'"[19] The maps that would eventually be produced by the expeditions would allow for the Labrador summer fishery to be extended more safely into heretofore uncharted waters and would eventually be made to cohere through the mission's practices of infrastructural mediation and their outlet in the building and maintenance of extractive networks for the fisherfolk prosecuting the fishery.

In many respects, the narrative of the mapping of the North Coast of Labrador demonstrates how the mission's infrastructural work would inevitably exceed its sphere of influence in northern Newfoundland and Labrador—relationships that seemingly localized infrastructural zones create can be process-based and future-oriented. Mapping the fisherfolk's extractive oceanic environment would lead to an accounting of the situated ecological effects of atomic testing. Infrastructural mediation, in this regard, is predicated on the processual relationship building that the expedition itself performed and that so often instrumentalizes resource frontiers and extracts a nonhuman, ecologically inflected conception of vital life. Much like Parks's understanding of "vertical mediation" and the drone's functioning as a relational technology, oblique aerial photography was part of the mission's practices of infrastructural mediation as they reshaped both fisherfolk lives and their ambient environment. "The drone is as much a technology of inscription," Parks writes, "as it is a technology of sensing or representation."[20] Oblique aerial photography both shaped the legibility of the North Coast of Labrador as a site of experimental geographic knowledge production and extracted its utility in perfecting the forms of compression the technique would lay claim to.

The logistical details of the survey work required a number of preparatory stages both on arrival in Labrador and prior to their departure. In order to be put into this infrastructural relationship, the North Coast of Labrador had to be made a part of the expedition's technical, and partly atmospheric, milieu. As noted above, the expedition bought a used monoplane from the Fairchild Aviation Corporation. Under the direction of Sherman Fairchild, whom Forbes had consulted throughout the lead-up to the expedition's departure, the company was a pioneer in the civil application of aerial photography. Fairchild was in fact behind a group of companies, including Fairchild Aerial Surveys, that was now capitalizing on the adoption of aerial reconnaissance practices and technologies deployed during World War I for the purposes of mapping growing urban areas from the air.[21]

With increased metropolitan densities came the need for more accurate land management. Fairchild aerial maps of New York City made in 1921 and 1924 would become documentary touchstones for this growing industry. Aerial photography and the mapping process of overlapping mosaics could provide the most accurate description of these newly contested tracts of purchasable land.[22] The plane that the expedition acquired had previously served the Bell

Telephone Company as a flying laboratory that undertook research into the development of aircraft radio equipment.[23] As the *Ramah* made its way by water, the Fairchild seaplane, piloted by Harold G. Crowley, followed in fits and starts, arriving in some locations hours before the boat. In addition to the Fairchild seaplane, Forbes had James K. Brownwell and Charles J. Hubbard fly his own open-cockpit Waco biplane to Labrador. The presence of the Waco would serve as a safety precaution, in case any mechanical issues arose with the Fairchild, as well as allow Miller and others to take short reconnaissance flights along the coast. Due to its high cost, the Model F-4 surveying camera was rented from the Fairchild Company. Moreover, the company provided a specially trained photographer, Sydney O. Bonnick, to operate it.

On the ground in Seaplane Cove, the members of the expedition set to work establishing the necessary markers for the surveying project. The prime marker for the entire survey would be an automatic tide gauge, loaned by the US Coast Guard and Geodetic Survey and installed at the foot of a steep shore of Seaplane Cove. The tide gauge marked the mean sea level, which, in turn, "served as the datum for all altitude measurements of the entire area."[24] By July 23 the crew, led by Hogg the architect, had also completed the building of a photographic darkroom next to a brook close to the *Ramah*'s anchorage. The darkroom had to be placed near a body of freshwater, in order to wash the film. The film itself, at 100 feet by 9 inches, was large and plentiful, and so the darkroom had to have sufficient drying racks, which were protected both from the elements, with a roof, and from the ever-present mosquitoes, with netting, as they otherwise would have left their footprints on the drying film.[25]

With the majority of the expedition's members preparing the survey's groundwork in the vicinity of Seaplane Cove, Forbes, Crowley, and Miller could look ahead to the actual aerial photography that would take place over the next several weeks. Given the unpredictable weather that predominates on the North Coast of Labrador in July and August, Forbes would constantly be watching and measuring various climactic variables in order to determine when optimal flying conditions would prevail. While the Fairchild plane could actually cover a considerable territory in a relatively short period of time, it was rather the work undertaken by the photographer and the camera that was more difficult to both maintain and coordinate. Part of the novelty of Miller's method of planetabling was to use the obliqueness of the photographs to project the cartographer's eye farther, pushing back the horizon line in order to encompass more land in each photograph. As he explains in a *Geographical Review* article published just prior to the expedition's departure

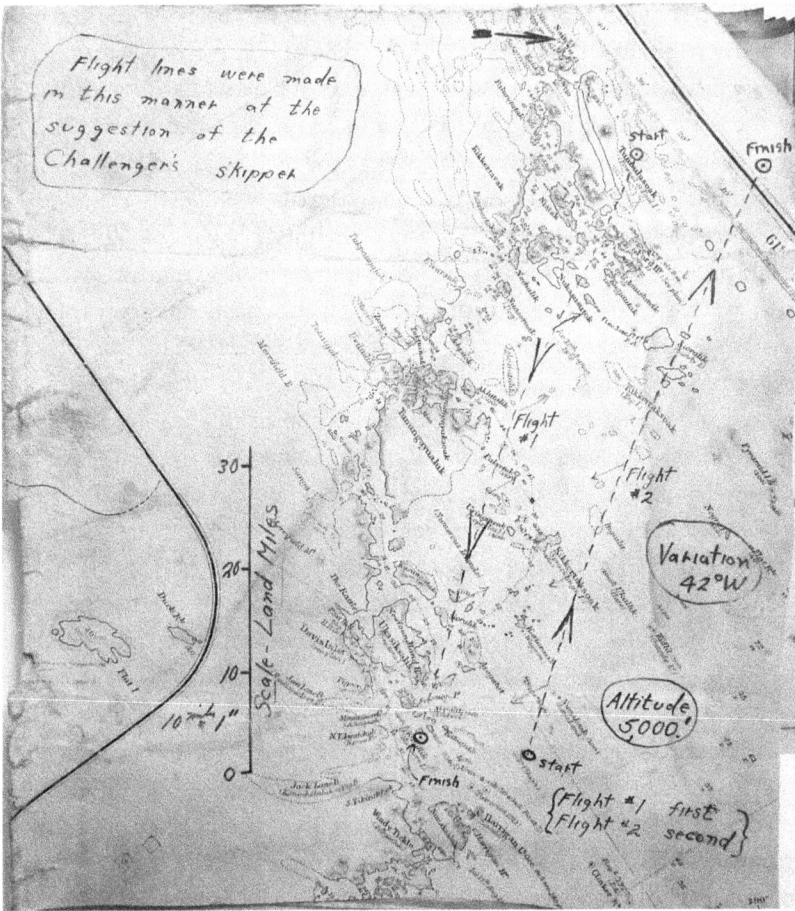

Flight lines were made in this manner at the suggestion of the Challenger's skipper

Start   Finish

Flight #1

Flight #2

Variation 42°W

Scale - Land Miles

30 —
20 —
10 —
0 —

Altitude 5000'

Finish   start

{Flight #1 first }
{Flight #2 second}

FIGURE 3.4. Map showing some of the flight lines of the expeditions from the 1930s. Box 100, folder 2195, Labrador Expedition, 1932, Correspondence Oct.–Dec. 1932, Alexander Forbes Papers, Francis A. Countway Library of Medicine, Center for the History of Medicine, Harvard Medical Library.

in April 1931: "The method developed in this paper depends on the assumption that it is possible to determine the horizontal on the photograph by trial and error, provided the silhouette of the landscape against the sky is visible. It enables the position of the air station in space and the tilt and direction of the camera axis at the time of exposure to be determined very quickly by graphical methods and by simple formulae depending on the fundamental principles of perspective."[26]

This technique of oblique aerial photography had been used in the past, mostly by Canadian government surveyors, notably Surveyor General E. Deville, who would use it to map vast if flat areas of land.[27] Miller's contribution to the production of small-scale contour maps was, as noted above, really on the level of economic efficiency as well as the application of the technique to mountainous terrain. In comparison to the costly production of maps using conventional ground- or air-surveying methods, oblique aerial surveying could produce a map at roughly one-third the cost and in a much shorter time period. The technique was starting to find favor by enabling surveys of more marginal and less instrumentally manageable territories. This was a crucial characteristic that made the North Coast of Labrador, led by Grenfell's suggestive question directed to Forbes, into an appropriate testing ground for the advantages presented by planetabling from the air; it was also a trait that merely extended long-established and largely, if not exclusively, British imperial networks of communication that would connect metropolitan interests to far-flung spaces of resource appropriation.[28]

Miller's original calculations for the method stemmed from an Antarctic expedition led by Richard Byrd, the famed American naval officer and explorer, from 1928 to 1930. In a memorandum to the expedition, Miller described a "field procedure . . . which would enable photographs taken during an exploratory flight to be utilized as material for a map without adding unduly to the cost of the expedition."[29] As with previous uses of oblique aerial photography, particularly by the Survey of India, it had hitherto been limited to reconnaissance work rather than to the more detailed needs of actual mapping. Miller also had some experience using the method in mapping remote areas of Peru.

The method itself relied on trigonometrical calculations that read several planes in relation to each other: "In developing the method and formulae for finding the horizontal position of the air station, its height above the datum level and the orientation and tilt of the camera axis to the horizontal it is necessary to consider three reference planes. These are the plane of the photograph, the horizontal map plane, and the principal plane of the camera, which last is defined as the vertical plane containing the optical axis."[30] In this system, the line of flight of the plane, while not all determining, is important in establishing what horizon of territory is to be covered. As Miller notes in his article, rather than take overlapping mosaic photographs as a sequence of parallel images, oblique aerial photography allows for panoramic coverage along the plane's flight line. In Miller's opinion, the best way of capturing the horizon line with a single-lens camera would have it pivot at set intervals and at a precise and consistent angle, a method that, he notes, a multi-lens camera could achieve at higher cost if more efficiently.

On July 24 the prevailing fog began to lift, and Miller could go about choosing the ground stations for his triangulation. He selected three in the vicinity of Seaplane Cove: one on the top of a nearby headland, Nehungatelek; a second at the top of Mount Roundabout; and a third at a height of two thousand feet on a ridge leading to Mount Tetragona.[31] To the southeast of the cove, in a broad and flat tundra, Miller established the survey's baseline, the accurate measurement of which, if properly aligned with fixed astronomical points, would provide the scale of the survey. The baseline was measured three times and was determined to be two thousand feet long. The positioning of the three ground control stations was also predicated on the ability to triangulate with the flag of the tide gauge from at least two of them at once and, thus, according to Forbes, "fix the geographical position of the entire system."[32]

In addition to the strictures of the photographic technique and weather, other variables, such as the expedition's limited supply of airplane fuel, went into the calculation of flight times. With clear skies on the afternoon of July 26, the aerial photography began. Crowley, Bonnick, and Smith took to the air, and during a one hour and forty-five minute flight from Nachvak Fjord to Eclipse Harbor, a point beyond the halfway mark to Cape Chidley, forty oblique views were taken at an altitude of seventy-five-hundred feet. While the expedition would stay on the coast of Labrador for over a month, the heavily weather dependent aerial photography would actually be accomplished in a matter of (nonconsecutive) days. On August 3, deemed by Forbes to be "the greatest day of the expedition," the Fairchild was in the air for a total of seven hours, with 330 pictures taken.[33]

By the end of the expedition, more than six hundred negatives were developed. As Miller notes in his contribution to *Northernmost Labrador Mapped from the Air*, this shift in the use of what had hitherto been exploratory aviation and exploratory photography—more or less the equivalent of a quick jaunt and sketch—for precise mapping work marked an important moment in how mapping expeditions would be organized in the 1930s. "Exploration by air had become the fashion," Miller writes, "and it seemed a pity that the photographic material available—mostly high obliques—could not be used for the purpose of map sketching."[34] The instrumental use of the photographs depended on the accurate translation of their topographical data onto a two-dimensional map. This was precisely what Miller's method achieved, and it was in Labrador in the summer of 1931 that he gathered the raw map material that would occupy much of his time over the next six years. Those environmental markers that at the outset of the expedition grounded the photographs, a medium of capture to be translated into other media forms, in the mean sea level and peaks of the North Coast, were the interfaces that linked the territorial conditions of northern Labrador to an infrastructural relation that could make legible the utility of such colonial spaces of inscription. In being a largely unmapped settler margin, the North Coast of Labrador inhabited a field of infrastructural mediation that challenged, as Innis and others have observed, established practices of colonial administration and, as the narrative of oblique aerial photography bears out, media creation.[35] "Colonial space is not simply acquired," as Jody Berland writes; "it is not an object. Its usable topographies are shaped, in dialectical interaction with its own agents and resources, to serve the requirements of empire."[36]

After the conclusion of the 1931 expedition, the majority of Miller's time would be spent working over the photographs in the New York offices of the AGS. Once the negatives were developed and Miller obtained a comprehensive idea of what parts of the coast and interior had been covered, he quickly realized that a second and possibly third expedition would have to be made to cover the missing sections of the territory to be mapped; across the medium of capture of high oblique aerial photography he encountered an environmental resistance to his infrastructure relationship building. Such impeding factors as the random spacing of the flight lines and altitudes, the prevalence of long and deep shadows in many of the photographs taken later in the day, the need for a higher average flying height in the range of three thousand meters, and the tilt angle of the camera, which was less than 10 degrees but would have greatly increased the area covered in each photograph had it been at an average of more than 15 degrees, all of these contributed to the need for additional aerial photographs of targeted areas of the coastline and interior. As so many other Grenfell Mission practices of infrastructural mediation show, there are often lingering, ecological frictions between a given environment and its being set in an infrastructural relationship—clouds that obscure horizon lines, fog that keeps ships moored, and sea ice that prevents

FIGURES 3.5 AND 3.6. (*facing page*) Oblique aerial photographs of the North Coast of Labrador, with base and tide gauge station marked in the lower photograph.
Box 100, folder 2187, Alexander Forbes Papers, Francis A. Countway Library of Medicine, Center for the History of Medicine, Harvard Medical Library.

human access. The North Coast of Labrador was indeed a "milieu" in Georges Canguilhem's sense, one that was a complex (and inhabited) ecology, like his example of the trade winds, that maintained a vitality independent of its colonial administration: "Trade winds displace surface seawater warmed through contact with the air, deeper cold waters rise to the surface and cool the atmosphere, the low temperatures lead to low pressures, which give rise to winds, and the cycle is closed and begins again."[37] Rivers, mountainous shores, salmon runs, these too were all part of Innu and Inuit living land.[38] In being taken up into the mission's practices of infrastructural mediation, the North Coast of Labrador became a "geographical factor" predicated on infrastructure relationship building—what Canguilhem calls a "geographical configuration" anchored by settler intention.[39]

In order to extend this need for further geographical knowledge production, both the 1932 and 1935 expeditions were quickly organized yet undertaken at a much smaller scale than that of 1931. Forbes only made the return trip in 1935, with this final expedition having as its purpose to photograph very particular sections of the coastline where shapes were hidden behind deep fjords or where the camera had not been fully trained on mountain chasms. A secondary purpose of the expedition was to determine possible harbors of refuge in the Cape Chidley region for the increasing number of cargo ships making their way to Europe from the port of Churchill, Manitoba. By 1934 Miller had become better acquainted with the scope of the photographs and so was able to determine with greater accuracy the absolutely essential spots that Forbes and Hubbard, his flying companion, had to cover. Forbes, using a loaned 5 × 7 Fairchild camera, put to use the relatively new capabilities of Dufaycolor film. One of the mission's backers, and a possible early candidate to take on the piloting duties for the 1931 expedition, was Major Sidney Cotton, who, incidentally, was one of the founding partners behind the new film venture.

"The true test of the method," Miller writes in *Northernmost Labrador Mapped from the Air*, "is in respect to the speed and cost of making the map."[40] Given that his use of oblique aerial photography was in its early, experimental stages for the mapping of mountainous terrain, there were numerous and inevitable delays in the production of the final maps, which appeared as a supplement of six sheets placed within the 1938 book. Miller realized early on that, for the sake of efficiency, he would need a new measuring instrument that could increase the speed of resection and plotting by measuring angles directly from the photographs themselves. Miller drew up the plans and contracted the Mann Instrument Company of Cambridge, Massachusetts, to build it, and the final instrument was in use at the AGS's office in March 1933.[41] That same year, Miller was sent by the AGS to assist the Louise A. Boyd expedition to East Greenland, leaving only Lieutenant G. R. Phelan, loaned by the US Hydrographic Office to assist in developing the new method, and Miller's assistant, John Kay, to continue resecting the photographs and proceeding by a sort of calculated trial and error. By 1934 another instrument, a single eyepiece plotter capable of drawing detailed sections of coastline directly from the photographs, was to be designed and built by the Mann Instrument Company for use in 1935. By Miller's estimation, the final topographical maps on the scale of 1:100,000 were completed at a rate of roughly 850 square kilometers per man per year. Given the minimal cost, in both time and capital, of the initial ground survey work and that of procuring and developing the film, which ultimately would come out to be more than that of chartering the Fairchild plane, along with the mere days of aerial photography it took to cover the necessary territory, Miller's calculus was that his method of oblique aerial photography, if shifted out of its experimental phase, could yield the accurate mapping of mountainous territory to a scale of 1:100,000 at a rate of better than a thousand square kilometers per man per year.[42]

Miller's calculations land on the infrastructural relationship yielded by the North Coast of Labrador, one that could couple a quantifiable compression of time, space, and capital with analogic "mountainous territories" yet to be mapped. Miller's descriptive work is thus also a means of drawing out the temporal and spatial aspects of infrastructural mediation; how it is a description of an original vital milieu that gets characterized and, here, compressed into an infrastructural set of relationships that bear out how settler colonial mapping practices were "acts and processes of temporarily stabilizing the world into media, agents, relations, and networks."[43] Rather than take Miller's calculation (1:100,000, i.e., a thousand square kilometers per man per year) as a definitive statement and accurate means of representation, infrastructural mediation sets it into Kember and Zylinska's processual and emergent understanding of a vital milieu that possesses an agency that can destabilize practices of temporal and spatial compression. The North Coast of Labrador, a thus far marginalized part of the fisherfolk's resource frontier, could talk back by resisting settler forms of environmental characterization and extraction through its own ecological recalcitrance made up of mountainous terrain and steep, shadow-filled fjords.

Indeed, the increase in speed and the concomitant adequacy of the final map's accuracy only partly explain the need for the return expeditions of 1932 and 1935. In his "Ground Survey Report," written immediately follow-

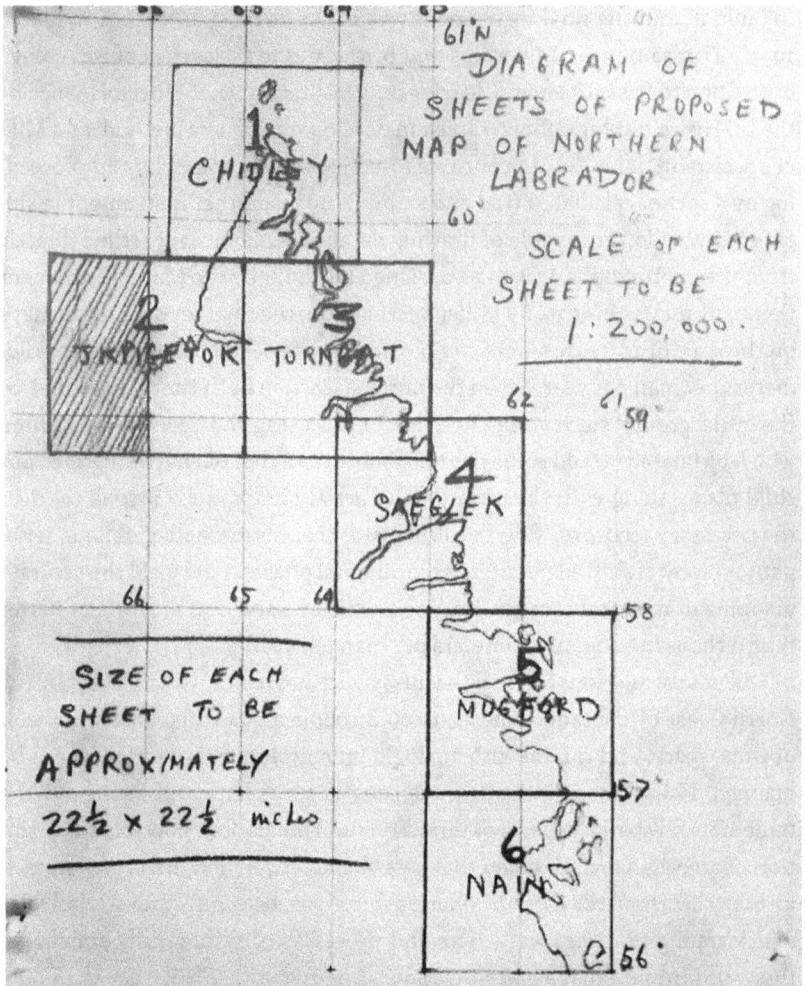

FIGURE 3.7. "Diagram of sheets of proposed map of Northern Labrador," in O. M. Miller's ground survey report. Box 100, folder 2187, Alexander Forbes Papers, Francis A. Countway Library of Medicine, Center for the History of Medicine, Harvard Medical Library.

ing the 1931 expedition, Miller notes: "Both as regards the extent and accuracy of the work, the survey was handicapped by the unavoidable shortness of the time available and the uncertainly of the weather conditions within that time. This latter factor had the psychological effect of forcing the pace with the result that the careful reconnaissance and planning so desirable in trigonometrical work was for the most part absent."[44] This was a recalibration of Miller's equation that would take into account the gains and losses brought about by speed in surveying work. The six years it would take to process the photographs were an aftereffect, a harnessing, of this acceleration of scientific advance. The intersection of civilian aerial photography, aerial exploration, and experimental surveying established a novel means of representing and interpreting the North Coast of Labrador as both navigable and accessible.

Oblique aerial photography, while reliant on high oblique photographs taken on flight lines at precise angles, was above all else a medium of compression. The photographs were a collection of planes to be interpreted and projected onto the two-dimensional surface of a map—processual dimensions of infrastructural mediation that highlight its ties to photography as a medium of compression that had to bind together the vital North Coast of Labrador and Miller and Forbes's establishment of oblique aerial photography as a legitimate mode of geographical knowledge production. It is a narrative that echoes what Jonathan Sterne sees as the interrelationships between the histories of communication and media of compression. "Starting from compression, communication has a 'network reality,'" Sterne writes. "This is to say that it is not a binary relationship between sender and receiver mediated by a medium but rather an ensemble of relations that only produce the moments of transmission and reception after the fact."[45]

Sterne works out from Gilbert Simondon's positing of a "relational causality" in On the Mode of Existence of Technical Objects, wherein a relation precedes phenomena on which it has effects: "Elements that materially are to constitute the technical object, and that are independent one of the other, lacking an associated milieu that precedes the constitution of the technical object, must be organized in relation to one another by means of circular causality which will exist once the object is constituted. What is involved here, then, is a conditioning of the present by the future, or by what up to now does not exist."[46] Simondon deploys the example of the Guimbal dam in the Philippines to describe how "the water moved by the turbine cools the turbine, enabling it to work at the properly regulated temperature to move water."[47]

As the submerged narratives underlying the genesis of oblique aerial photography as a medium of compression reveal, it was precisely an infrastructural "ensemble of relations" that would only "produce the moments of transmission and reception after the fact" in the form of the maps that were included in *Northernmost Labrador Mapped from the Air*.

Sterne's observations from the histories of communication and compression align with how the mission's practices of infrastructural mediation were predicated on shifting milieus and such a meta incognita as the North Coast of Labrador into a "circular causality" that could set them into an open-ended infrastructural relationship while also negating their independent ontological vitality—how water is first set apart from a turbine. As a "network reality" in this history of oblique aerial photography as a medium of compression, Forbes's 1938 book signals one evolving node in the experimental technique's settler-civilian-to-military application. Through Simondon, Sterne highlights how infrastructural mediation necessitates the negotiation of compression in the production of colonial territoriality and temporality: the mission's aims of both increased and safer access for the fisherfolk and Miller's development of a time-based form of geographical knowledge production that could circumvent established relationships of map-based compression. The North Coast of Labrador, in the eyes of Forbes and Miller, was a matter of definition—a poorly known meta incognita that had to be set into the media-derived terms provided by oblique aerial photography. "Whether in its audio or data varieties, compression accommodates signals to infrastructure," Sterne writes. "But it also transforms infrastructures by enabling them to carry different kinds of signals."[48] The North Coast of Labrador was made to become a signal that shaped the infrastructural relationship building enacted by the development of oblique aerial photography. As the publishers proofs of *Northernmost Labrador Mapped from the Air* suggest, this was an unbinding of settler logics of appropriation and their originary sites of dispossession: where Seaplane Cove was made.

FIGURES 3.8–3.11. (*facing page*) Publishers proofs of images to appear in *Northernmost Labrador Mapped from the Air*. Box 45, folder 1552, Re: "Northernmost Labrador Mapped from the Air," 1938 (no. 17 of 23 folders), Alexander Forbes Papers, Francis A. Countway Library of Medicine, Center for the History of Medicine, Harvard Medical Library.

Operation Crossroads

With the publication of the expedition's results in 1938, the logistics of access on the North Coast of Labrador did not drastically change. Rather, as I touched on in the previous section, the implementation of Miller's technique became an emerging locus classicus in the fields of aerial photography and aerial surveying.[49] It was this sanctioning of a new technical practice that the expedition's maps ultimately charted. The North Coast of Labrador was the ever-convenient territory, or, following Canguilhem, discounted milieu that had to stand in for futural forms of environmental characterization. Forbes, writing in 1934 to Isaiah Bowman, the head of the AGS, would state, "Ever since I learned th[e] significance of Miller's new method, I have felt that our most important objective was the introduction of a new method rather than the survey of a particular territory. Recent developments seem to bear out that view."[50]

After Forbes's flight to Cape Chidley in 1935,[51] he continued to foster an abiding interest in aerial surveying. At the onset of World War II, at the age of fifty-eight, Forbes rejoined the navy as a lieutenant commander and over the course of the war was assigned to several posts. His first, drawing on his medical expertise, consisted of testing electroencephalography as a means of determining the fitness of airplane pilots. The second, and more long lasting, was as a technical adviser to a reconnaissance expedition to Labrador headed by Captain Elliott Roosevelt (one of the president's sons). The purpose of the expedition was to select airfields in order to allow fighter planes to bypass the dangerous North Atlantic crossing by ship and fly, via Iceland, to the European theater of war. Up until 1943, Forbes would undertake various forms of survey work, both hydrographic and land-based, for the navy, first at Koksoak River Station, then near Frobisher Bay, with the aim of amassing a great deal of territorial data that ultimately would determine the strategic utility of Labrador for the American war effort. Much of this experience would go into Forbes's 1953 publication, *Quest for a Northern Air Route*.[52]

Many of the airport sites and water landings that Forbes identified during his surveying work would not end up being used for air travel. Instead, they became the first infrastructure for a weather-monitoring system that would stretch across the Atlantic.[53] Again, infrastructural mediation is set into the process of infrastructure relationship building, with the vital milieu of Labrador being made to accommodate colonial forms of spatial and temporal administration. As Parks and Starosielski ask: "What kinds of media distribution do these 'natural' environments support? How are nonhuman forms of life affected by the presence of media infrastructures?"[54] The North Coast of Labrador served as an environmental medium of compression that, in turn, could shape the networked production of colonial and military control. The shadows cast by the infrastructural relations that laid claim to Labrador, from mapping to weather forecasting, served to obscure and delay Indigenous claims to territorial sovereignty—the Nunatsiavut (Our Beautiful Land) in a then-distant future.[55]

The war effort was also instrumental in allowing Forbes to make the transition from civil amateur to military expert in the field of aerial surveying. By 1944 Forbes would be able to submit the manuscript "Short-Cuts in Long-Distance Photogrammetry" for publication in the American Society of Photogrammetry's journal, *Photogrammetric Engineering*.[56] While the manuscript notes how aerial photography has most often been used in applications requiring high precision, it also remarks on how the observation and documentation of such natural phenomena as cloud formations over the ocean or the distribution of floating material on the surface of water bodies, "things of interest to aerologists and oceanographers,"[57] can be facilitated by the technique. Subsequently, in February 1945, the navy's Hydrographic Office and Photographic Intelligence Center published the manual "Introduction to Oblique Photogrammetry." While ostensibly an anonymous, institutional text, "prepared for the purpose of acquainting Navy personnel engaged in aerial photographic work and mapping with the elementary phases of oblique photogrammetry,"[58] Forbes's correspondence reveals that he was its principal author. "The scope of this manual is limited to the use of high obliques," Forbes writes in the introduction, "and especial emphasis is given to rapid methods designed to give approximate results in situations in which speed is more important than precision."[59] This popular militarization of parts of Miller's technique is remarkable not only because it shows the extent to which Forbes learned the basic tenets of the surveying work but also because it reinforces Labrador's position as an infrastructural territory caught in an evolving set of relationships predicated on its topographical attributes.

In a communiqué between the Hydrographic Office and the Photo Interpretation School at the US Naval Station in Washington, DC, the guiding rationale behind the manual is laid out:

The Photogrammetric Unit of the Amphibious Training Base at Camp Bradford, Norfolk, Virginia, has recently encountered situations in which rapid use must be made of miscellaneous oblique photographs, ill-suited to standard photogrammetric procedures, for making maps of vital military areas. Recent experience has suggested to the officers of this unit the need of disseminating to photo interpretation officers in the

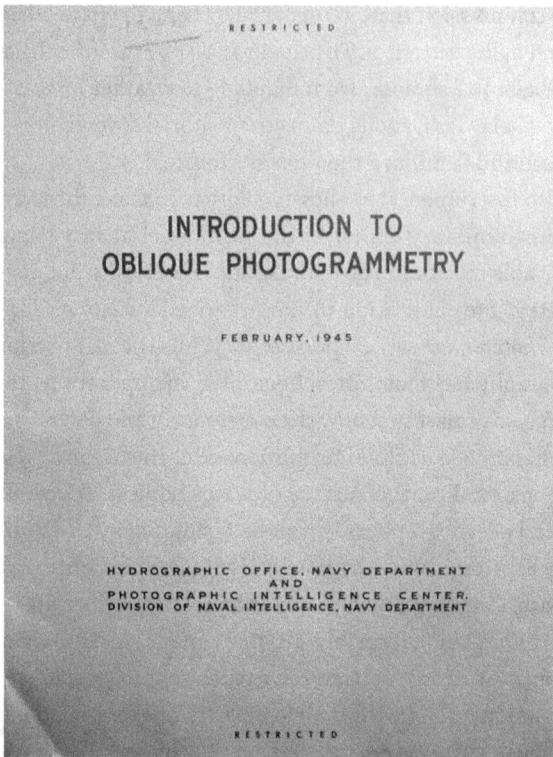

FIGURE 3.12. Title page of "Introduction to Oblique Photogrammetry," Hydrographic Office and Photographic Intelligence Center, Division of Naval Intelligence, Navy Department, February 1945. Box 46, folder 1579, Re: "Introduction to Oblique Photogrammetry," 1945 (no. 5 of 5 folders), Alexander Forbes Papers, Francis A. Countway Library of Medicine, Center for the History of Medicine, Harvard Medical Library.

war zone a manual outlining simply the most modern methods of using oblique aerial photographs for mapping situations calling for speed rather than precision.[60]

At the suggestion of Lieutenant Colonel Charles H. Cox, officer-in-charge of the Photographic Interpretation Center, the manual was also to be included as a supplement to the "Photographic Interpretation Handbook of the United States Forces."[61] The echoes to the 1931, 1932, and 1935 expeditions are numerous, namely in the manual's emphasis on the conjunction of speed and precision with an economy of means (here, photographic material and time rather than capital). The specific forms of local knowledge that Forbes took away from those mapping expeditions to Labrador are here made evident in their translation into a stripped-down accounting of what sorts of scientific surveying advances they enabled—practices of infrastructural mediation

BASIC PRINCIPLES

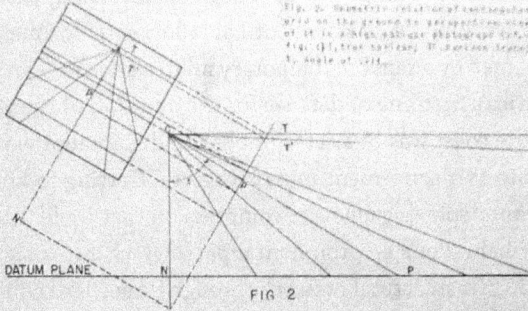

FIGURE 3.13. "Basic Principles," in "Introduction to Oblique Photogrammetry," 2, 3. Box 46, folder 1579, Re: "Introduction to Oblique Photogrammetry," 1945 (no. 5 of 5 folders), Alexander Forbes Papers, Francis A. Countway Library of Medicine, Center for the History of Medicine, Harvard Medical Library.

that reverberate along militarizing infrastructural networks. Newly repurposed for military use, they show how the multiple narratives surrounding the aims of the expedition would continue to evolve as the practices, and what could be thought of as the infrastructural outcomes of the Labrador expedition, continued to unfold into broader geopolitical arenas.

What role did the Grenfell Mission hold in facilitating this shift from settler-civilian to military use? This is perhaps too essentializing a question, for the mission acted more as an institutional conduit—inviting the initial survey, providing basic logistical support, and, upon publication of its results, fostering a sense of collective ownership and "advance" around their work that would make the North Coast more available for readier and safer forms of maritime access. Yet, while ownership of the photographs and the maps resided with the AGS, the technical ends of high oblique aerial photography as a scientific practice could be taken up by Forbes and repurposed through both self-teaching and citation (Miller figures in the manual's bibliography). The tangible benefits from the mapping expeditions that would ultimately accrue to the settler fisherfolk working off the coast of Labrador were not merely negligible but were enframed by a sense of temporary militaristic use—an occupation of North America's first line of defense for the duration of the war. Forbes's knowledge of the coast was of a cartographic kind. Labrador would thus continue to circulate as a convenient meta incognita. Getting to know this unknown shore meant deploying media of compression that could translate the North Coast of Labrador's environmental parameters into a signal that could create its own infrastructural network through its settler-civilian-to-military circulation.

Nearing the end of World War II, Forbes was detailed to Operation Crossroads. With the Bikini Atoll as a broad field of experimental operations, Forbes was charged with mapping and measuring the waves generated by the atomic tests. Given a "ring-side view" from May to July 1945, Forbes applied his mapping techniques to assess an important, if marginal, outcome of the atom bomb.[62] In a 1946 letter to the editor of the *Washington Post*, Forbes takes issue with the paper's editorial characterization of the experiments at Bikini as "'not tests in any real scientific sense.'"[63] Rather, Forbes saw the Crossroads project as being "the most colossal, coordinated multiple scientific experiment in all history."[64] While reluctant to apply any qualifiers to the tests, he was adamant in situating them in a scientistic ethical framework that could bracket "whether or not it was right to shorten the war and save countless thousands of American soldiers in that way."[65] This epistemology of scientific expediency, of making Bikini yet another available meta incognita, could make the collective existential threat of the atomic bomb into a worthwhile opportunity for scientific advance—infrastructural mediation scaled up to the planet as a whole. "When the data have been analyzed," Forbes writes, "they will constitute fundamental contributions to physics, chemistry, geology, seismology, hydraulics, oceanography, aerology, marine biology and mammalian pathology."[66] By the end of World War II, the North Coast of Labrador may have come to be mapped, but it certainly had not become known. Like most waves, scientific progress came in and receded, leaving in its wake a precision-mapped shoreline newly opened to the world of geopolitics.

FIGURES 3.14–3.16. (*above and following pages*) Oblique aerial photographs of nuclear test at Bikini Atoll. Box 112, folder 2386, World War II Photos #91: 51–92 . . . 183: 52–87, Re: Bikini, Alexander Forbes Papers, Francis A. Countway Library of Medicine, Center for the History of Medicine, Harvard Medical Library.

* * *

SLOW DISTURBANCE, "CHANNEL 12"

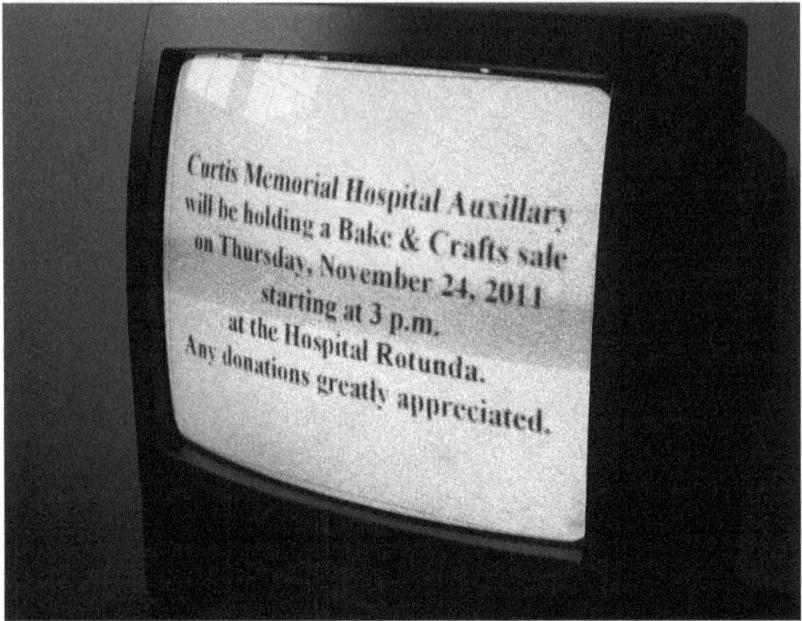

Announcements (*above and following pages*) broadcast over St. Anthony's public access channel. St. Anthony, Newfoundland and Labrador, 2011. Photographs by author.

St. Mary's ACW
will be holding a Craft/Bake sale
and afternoon tea on Sat. Dec. 3
at 2 p.m. in the Church Hall.
All are welcome.
At this time we would like to thank
you for your support over the past year.
We would also like to wish you all a
Very Merry Christmas and a
Happy & Healthy New Year.

CHANNEL 12
YOUR COMMUNITY CHANNEL
BROUGHT TO YOU BY
Persona CABLE TV AND THE
AURORA NORDIC SKI CLUB.
WORKING FOR YOU, ALL DAY,
EVERY DAY.

4. THE PROMISE OF EXTRACTION

In March 1931, a sealing ship, the SS *Viking*, set off from St. John's destined for the unstable, if expansive, ice fields off the coast of Labrador. Among the working sealers was a group of American filmmakers. The screenplay still known as *White Thunder* had been shot the previous year in Quidi Vidi by Varick Frissell, an aspiring director and, beginning in the 1920s, a volunteer with the Grenfell Mission. Over the course of the decade, Frissell, a recent Yale graduate, would make several documentary films on Labrador, including *The Lure of Labrador* (1928) and *The Great Arctic Seal Hunt* (1928), both inspired by the work of Robert Flaherty.[1] Frissell's screenplay told the story of the rivalry between two sealers for a woman's love and their adventures while on the hunt. With the basic scenes of the film covered in 1930, Frissell deemed it necessary to supplement these with realistic footage of the seal hunt, action sequences of a kind that would be able to bring the realities of the shifting land of laboring sealers to film house audiences across the world.

On March 15, as the ship was immobilized by pack ice near the Horse Isles, an explosion in the *Viking*'s powder room killed twenty-seven men: sealers, film personnel, and crew.[2] *The Viking*, as the film would ultimately come to be called, was released in the summer of 1931 to audiences in New York, Toronto, London, and Paris. It was the first Hollywood-style sound film made in Canada; it is also a landmark disaster in film history.[3] The film was funded by Paramount Pictures, and by many standards of the time it was

FIGURES 4.1–4.10. *(following pages)* Stills from the filming of *The Viking* (1931), directed by Varick Frissell. Labrador Institute Archives, Labrador Institute, Memorial University.

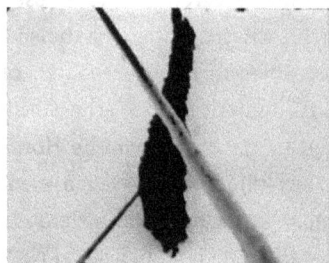

an odd combination of documentary-style reportage (with actual sealers filmed while on the hunt) and amorous filmic fiction. Yet it was precisely the specifics of the geographic locale that had inspired Frissell to set his somewhat improbable scenario off the coasts of northern Newfoundland and Labrador. The scenes of sealers in *The Viking* make for mesmerizing viewing. They show the roll of the ocean and its crusty, shifting pans of ice, with lightly clad sealers jumping with their icepicks from one to the next, sometimes falling into the North Atlantic waters and pulling themselves up and out. If it would seem improbable to have a film crew filming on rolling ice, it is all the more disconcerting to be at that next remove, watching the crew film the sealers (and actors) as they race across the pans, shooting a somewhat slow if dangerous chase scene. The footage is grainy and lurches here and there. The *Viking* (the ship) carried explosives for precisely the situation in which it found itself—trapped in ice. A sequence in the footage shows an assembled line of men all holding a line attached to the ship, presumably in an attempt to free it through sheer human force. Frissell, after authenticity rather than studio theatrics, brought his camera and crew to the sealers' environment. The footage's oceanic roll is a material reminder of the sites where medium and environment merge.

THE LABRADOR INSTITUTE, a division of Memorial University in Happy Valley-Goose Bay, holds a collection of never-released film footage belonging to the IGA. Consisting of six distinct reels dating from the 1930s to the 1950s, it amounts to roughly two hours and thirty minutes of silent footage of the infrastructure (buildings, ships, roads) and social projects (plays, dog sled races, nursing) the Grenfell Mission created in the region at the apex of its influence. The viewer can silently travel along with one of the mission's medical cruises as it follows Labrador's coastline; witness the excitement and human flurry that accompanies the arrival of the summer steamer in St. Anthony; take in the joy of mission orphans at play, running races, and putting on elaborate pageants.

In the Grenfell Mission story, infrastructural mediation lands last on the medium of film. This is somewhat of a storied itinerary—a selective recasting of the varied practices of infrastructural mediation that the mission enacted. And it isn't one that is meant to discount how and why the mission privileged conventional media of representation. Far from it. Magic lantern slides, photography, and, eventually, film would all contribute to relaying the Grenfell Mission story of an outpost made up of

FIGURES 4.11 AND 4.12. Film stills from found footage of Grenfell Mission from the 1930s to the 1950s. Labrador Institute Archives, Labrador Institute, Memorial University.

needful and worthy Anglo-Saxon settlers feeding the empire. Across the mission's evangelical Protestant homiletic tradition, which I have thus far tied to its broader efforts at creating medical, financial, and territorial infrastructures of care and maintenance, images—particularly moving ones—often spoke not only louder than words but with much greater efficiency and clarity, as they could circulate independently of Dr. Grenfell's seasonal lecture tours.

The mission itself produced institutionally sanctioned and finished films. An article in the July 1928 edition of *Among the Deep Sea Fishers* states, "At last talking-moving pictures have been made of Sir Wilfred and the work

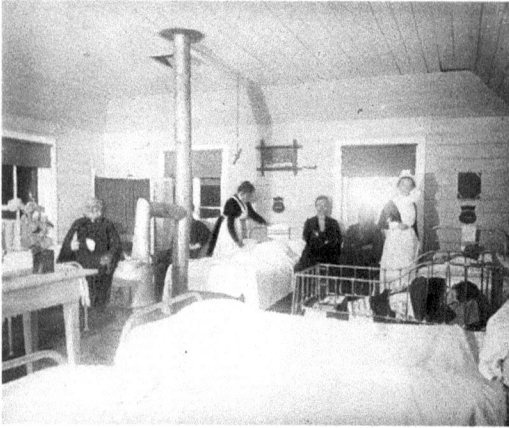

FIGURES 4.13–4.17. (*left and above*) "IGA Magic Lantern Shows." Collection of IGA magic lantern slides from the 1920s, Provincial Archives of Newfoundland and Labrador, accessed March 12, 2014, http://www.exhibits .therooms.ca/panl/exhibits/lanternslides.asp.

of the Grenfell Mission."[4] This film, assembled from nearly seven miles of footage, was largely shot by Frissell and Herbert Threlkeld-Edwards. While Threlkeld-Edwards was a mission volunteer keen to use his newly acquired DeVry machine,[5] Frissell, as I touched on above, was an aspiring director and documentarian.[6] It was the mission's first foray into the world of moving pictures and came at the significant cost of $3,000, which was spent mostly on preparation of the disparate footage into a series of vignette-like films by the Bristol Company of Waterbury, Connecticut. It was also, in the IGA's view, of the most value because it made up a permanent "living and imperishable record of Labrador's greatest personality."[7] Moreover, film constituted a kind of mechanical speaker that shifted the burden of public speaking away from Grenfell while also generating rental income from school, church, and university audiences interested in the mission's work. For the IGA, the medium of film became the best means of creating a lasting historical record of the mission's activities.[8]

Thus, while representing the mission's good works as well as the material needs of the impoverished fisherfolk was of instrumental concern for the mission, particularly with regard to its annual fundraising tours across the northeastern United States, Canada, Britain, and Ireland, it was a secondary priority in relation to the more urgent infrastructural work demanded by the fisherfolk's resource frontier—as I have addressed throughout, a settler colonial world that from the 1890s forward was characterized as in need of cyclical and profound reform. The settler infrastructure the mission helped build and maintain was equally an elemental infrastructure emerging from and grounded in the shallow, rocky soil and roiling North Atlantic coast its colonial infrastructural mediation was there to change. The representational work of the IGA footage lies in how it silently immerses the viewer in the mission's infrastructural reforms as processes of mediation. As I note in the introduction, environmental media studies can lay claim to a privileged purview of the emergent relationalities that can be mapped between vibrant ecologies and the material, media-related practices of ordering that coalesce in the very medium of this book. I have indeed, as expressed at the outset of *Slow Disturbance*, been assembling this minor and marginal story of the Grenfell Mission in order to articulate a very local enactment of practices of infrastructural mediation that are shaping anthropogenic relationships to extractive environments. Mediation is so often understood as a process that underlies all manner of narrowly circumscribed media technologies: a technical if not teleological channel that possesses a beginning and an end. Yet it is also a process that is constitutive of such environmentally inflected media as extractive resource frontiers that register, in this case, how a North Atlantic settler colonial project cyclically bound together infrastructures and environments in the service of sustaining an evangelical Protestant world. It is by telling a particular story, not only a marginal and minor but also an indeterminate one, that I am trying to scale down and render infrastructural the punctual, evanescent, and lively mediation that permeates and subtends all manner of lifeworlds, with a particular focus on those defined by the settler's promise of extraction.

Following Nixon's influential call for a stilling of the material relations that environmental justice demands, I have assembled the infrastructural story of the Grenfell Mission as one that can be made and unmade, cast and recast, in line with J. K. Gibson-Graham's understanding of "story" as a "performative ontological politics" that can trace the difficult relations that hold people, environments, and their material tensions together.[9] The Grenfell Mission story is one such performance of a media historiography that can both still and arrange the colonial lifeworld it was there to reform while also being anchored in northern Newfoundland and Labrador's near future and the relations to be traced between places like St. Anthony and the last fish of the North Atlantic. "Like any object of description," John Durham Peters writes, "the past is emergent."[10]

The relationalities that inhere across resource frontiers are acute in their articulations of environmental pasts that are frequently extracted in order to be forgotten. Much like the drying of cod in northern Newfoundland and Labrador was a process that could hold that past up for examination, the historiographical work *Slow Disturbance* performs is an attempt to still such processes of mediation not only to make out how the fish came first but also to create, following Katie King, a "transmedia story." For King, a transmedia story is a performative media ecology that can create knowledges through the ways in which it is constructed as an object of inquiry.[11] To some degree, this is a staging of a particular historiographical practice. Yet it is also a generative strategy that recognizes a hermeneutic hand in the configurations of historical artifactuality. In King's estimation, these sorts of strategies can contribute to "transdisciplinary knowledges" that work toward post-humanities approaches to agency (now distributed across multiple media platforms, fields of intentionality, and artifactualities) that recognize the limited control that writers (and their writing technologies) hold in the making of these agential environments.

I see environmental media historiography itself as a "slow disturbance landscape."[12] Relations between people and milieus, environments and infrastructures, can emerge and recede across storied itineraries that operate independently of a human subject. Scholars across this conjuncture can start to see how infrastructural stories are often settler stories too and that such a material politics of mediation is grounded not only in ecological situations but also in the durational dimensions that infrastructure building gives rise to—affects, dispositions toward the future, embodied understandings of the way it was. Environmental media studies and beyond can begin to situate the processes of mediation that can be found in the relations we build with responsive environments. This is not only an attunement to the "territorial archive" that van Wyck, following Innis and others, has graphed onto a storied understanding of multigenerational ethics—an essential supplement to the environmental effects argument, from carbon footprints to e-waste accumulation, prevalent in environmental media studies.[13] It is an orientation toward the ways in which we characterize stories as media that we deploy in the service of privileging certain lives over others; as Cameron has it, stories are indeed "material ordering practices" that have tangible effects on local lifeworlds.[14] Seeing environmental media historiography as akin to a slow disturbance landscape is a means to, following Kathleen Stewart as well, "fashion some form of address" that apprehends how central processes of mediation are to an expanded understanding of anthropogenic environment making.[15] Like a single film still, this book has set out to assemble one infrastructural story among others to show where mediation and environments merge and become legible through settler practices of extraction.

Within this slowed historiographical landscape, the hundred and fifty minutes of film spanning twenty years of mission influence give dimensionality to the durational aspects of infrastructural mediation and show how it was a missionary practice that occupied a tension between vital, environment-specific milieus and the material and ideological exigencies of settler colonial life. I have described how the fisherfolk's reformed resource frontier came into being as an open-ended process of infrastructural mediation that was predicated on the building of settler infrastructure—arrangements of a colonial world that were relational and, as the chapters in this book have borne out, tied to sited stories of emergence. Indeed, the media environments available for engagement to scholars working across media studies and allied fields are just such processes of mediation that emerge relationally and indeterminately through ecological frictions, spatial and temporal practices of ordering, and storied human experience.

The IGA footage, if it is in need of categorization at all beyond its inchoate historicity, is a case of indeterminate media. Indeterminate media are not only media that are difficult to categorize, but in a more materialist reading that exceeds characterizations of "remediation" or "residual media" as the interplay and co-shaping that occurs between old and new media forms, they designate media of capture that foreground *processes* of mediation and their ties to the durational dimensions of particular environments.[16] In this instance, the indeterminate medium of found footage captures the infrastructures the Grenfell Mission built, maintained, fundraised for, and circulated across northern Newfoundland and Labrador. It highlights how, over this twenty-year period, the appearance of docks, hospital boats, nursing stations, and hospitals, while representative of the mission's spatial and infrastructural administration of the fisherfolk's lifeworld, was also perhaps more consequentially a durational experiential horizon that their projection of settler infrastructure brought into being—a process of settler becoming that such resource frontiers foreground through their constantly unspooling processes of mediation: lives to be lived through infrastructure.

This characterization of the IGA footage demands a recognition, as Laura Mulvey suggests, of "cinema['s] . . . privileged relation to time, preserving the moment at which the image is registered, inscribing an unprecedented reality into its representation of the past."[17] Film's indexical quality registers and inscribes a moment of the past that "fixes a real image of reality across time."[18] This conception of duration, of bringing *filmically* real images of past realities into the present, gives film, and found footage in particular, a documentary valence that can foreground such practices of infrastructural mediation. These images become interpretable facts; as Mulvey points out, "the trace of the past in the present is a document, or a fact, that is preserved in but also bears witness to the elusive nature of reality and its representations"[19]—the indeterminacies between environment and medium, between infrastructure and environment that the mission's attempt at shaping what Berlant sees as the "living mediation" that so often inheres in infrastructural world-making projects. The IGA footage is both a filmic representation of an environment as well as a documentary, material embodiment of that environment: it foregrounds the relationships between singular instances of registration and the mission's practices of infrastructural mediation. The footage's articulation of what Mulvey calls "'film time,' the inscription of an image onto the still frames of celluloid," contributes to its occupying this gray zone of the indeterminate medium.[20]

By recasting relations between environmental conditions, registration, and duration, what does the footage make present? In large part it functions like a heuristic that indexes how the mission was a relational process of mediation predicated on its infrastructural work. Much like Peters's understanding of "infrastructuralism," "the call to make environments visible," the IGA footage can also be read as a media artifact belonging to the mission's practices of infrastructural mediation—not only a symbolic index but also a more markedly material inscription of the where and the what of that mediation. The footage is a medium environment, both indeterminate in its foregrounding of the durational, vital dimensions of missionary mediation

FIGURES 4.18–4.51. (*following pages*) Stills from Grenfell Mission found footage from the 1930s to the 1950s. Labrador Institute Archives, Labrador Institute, Memorial University.

"The Mission still goes on as he wished it. The things he did are still an inspiration to us to carry on. He helped us to help ourselves...."

yet also grounded, as Ursula Heise would have it, in the "material territories of built and natural environments,"[21] an infrastructural story *and* the "film time" of North Atlantic light on celluloid.

Yet the footage coexists with more rigidly administered ways of accounting for the representational stakes on which the mission relied. In the early 1930s several IGA board members and individual associations would put forward proposals for the continued development of film as the mission's privileged medium of representation. These ranged from those early "reels of pictures" made by the Bristol Talking Picture Corp. and directed by Varick Frissell, of "Sir Wilfred, in evening clothes, giving a brief history of the Mission, . . . pictures of the work on the Coast, a cruise of the Strathcona, a very good reel of Orphanage pictures, during which Sir Wilfred's voice is heard, explaining it all,"[22] to a possible children-oriented biographical film of Grenfell and the mission, to be directed by Mary Field, a pioneer in children's entertainment in Britain.[23]

The success of Frissell's first mission film led him to partner with his mentor, Robert Flaherty of *Nanook of the North* fame, to create the Labrador Film Company, based in New York. The production company, with the blessing of the Grenfell Mission, proposed to tell the story of a young doctor and nurse at one of the mission stations "based upon certain authentic and dramatic episodes taken from the history of Grenfell activities on the Labrador."[24] While the motion picture venture was a racialized commercial endeavor that ultimately did not come to fruition,[25] Frissell and Flaherty planned on donating thirty-five percent of the film's net profits to the mission while also claiming that the "cinema medium" would broadcast the mission's work to audiences around the world and thus generate interest in Grenfell's philanthropic projects of reform.[26]

One of the material, analog legacies of this forecasted cinematic future is the IGA footage itself—an outcome of missionary intervention that exceeds the bounds of its own institutional agency. It is material that can be taken up into the practices of ordering that shift across this interpretive slow disturbance landscape of infrastructural mediation itself. As Cameron stresses, "Stories are complex assemblages of people, places, and things; some may be narratively performed by humans, but they must be understood as relational networks of humans and non-humans, not as representations that somehow sit apart from the materials they represent."[27] The footage instigates a "mode of relation" that bypasses its own representational stakes and foregrounds the creation of an indeterminate media environment. "Rather than assuming to know in advance who and what matters in a given context," Cameron writes, "one proceeds by patiently tracing relations."[28] *Slow Disturbance* is one performance of the Grenfell Mission—a transmedia story assembled from its archival substances and the manifestations of living mediation that tie it to communities across northern Newfoundland and Labrador. The book is a single tracing of the present ineffability of colonial pasts with, as Stoler has it, "the narratives recounted about them, the unspoken distinctions they continue to 'cue,' the affective charges they reactivate, and the implicit 'lessons' they are mobilized to impart [that] are sometimes so ineffably threaded through the fabric of contemporary life forms they seem indiscernible as distinct effects, as if everywhere and nowhere at all."[29] Colonial ineffability can exceed representation too: it speaks of the vital mediation that the fisherfolk's resource frontier gave rise to and also of the material ordering practices that have tried to account for its emergence. Stories of infrastructural mediation don't always tend toward the sensible—sometimes indeterminate and open-ended, they always enact future relations across those tensions that bind them together as media artifacts. These are posthuman lessons that anticolonial and Indigenous scholars have long held up as reclamations for not only telling their own stories but also for creating more prominent and egalitarian epistemological spaces that foster their own materials, practices, and ways of making stories.[30] "The past, our stories local and global, the present, our communities, cultures, languages and social practices," Linda Tuhiwai Smith writes, "all may be spaces of marginalization, but they have also become spaces of resistance and hope."[31]

To view the IGA footage is to participate in the process of mediation that ties together the Grenfell Mission story, not between then and now, but rather in the diffuse, ineffable promise of extraction that permeates its indeterminacy. Stories of extraction are indeed material ordering practices. This way of describing the broad set of evolving relations that still and become temporarily concretized into a story is reminiscent of essential characterizations of infrastructure—as Berlant puts it, "the movement or patterning of social form."[32] *Slow Disturbance* has sought to trace the Grenfell Mission as a story of settler infrastructure. "Infrastructures reach across time," to again follow Cowen, "building uneven relations of the past into the future, cementing their persistence."[33] Ordering together and storying settler colonialism, particularly one of such long-standing in this subarctic milieu, and the mission's practices of infrastructural mediation to project their own real and imagined settler realities, is an attempt to characterize just one dimension of a scholarly orientation that works out from a position of settler accountability and that should find a more prominent place in environmental media studies. While the Grenfell Mission story tells how colonizers co-shaped their resource frontier, it is one whose infrastructural relations extend into the present. The promise of extraction is a story of persistence—of settler colonial relations that get inscribed across windswept, rocky harbors as docks, roads, and hospitals.

Spending time where infrastructure and environment meet is to bring into being the slow disturbance landscape of infrastructural mediation. Promises of extraction are stories for media studies scholars to tell. While I approach settler accountability, particularly in the sub-arctic and arctic, as a southern scholar who can mark and inscribe my own entanglements with such ineffable and persistent colonial relations, *Slow Disturbance* is nonetheless one storied relation in the anticolonial project of unsettling. "For settlers," Cameron writes, "*unsettling*—imaginatively, materially, politically—is an essential and ongoing task."[34] Practices of ordering, equally storied and infrastructural, can unsettle the mediations that are deployed in the project of resource frontier making in settler colonies. These arrangements foreground how the cross-generational promise of extraction is really an indeterminate one—like the IGA footage itself, a merging of infrastructural relationship building and milieu with no beginning, middle, or end—a settler lifeworld submerged in the present. What this book asks of the Grenfell Mission story is whose world is being sustained and through what relations? The lifeworld and afterlives of the fisherfolk of northern Newfoundland and Labrador show how infrastructural mediation is living mediation. Unsettling the promises of extraction they lived is to follow the enduring relations they built, maintained, and passed on—from first fish to last.

* * *

SLOW DISTURBANCE,
"SAMSUNG, HIGH-SPEED MECHANISM"

Viewing IGA found footage at the Labrador Institute. Happy Valley-Goose Bay,
Newfoundland and Labrador, 2011. Photograph by author.

Ernest Simms, a former mayor of St. Anthony, remembers looking out over the Newfoundland coast one night in the early 1980s.[1] Within the stipulated twelve-mile fishing limit, Simms could make out draggers and trawlers from around the world moving back and forth along the coast. In contravention of the regulatory framework that had been put into place, cod from the Grand Banks of Newfoundland were being fished to near extinction. Simms remembers how it was as though you were looking out over a city of lights with a moving skyline. Unbeknownst to the women and men of the province, this was the gradual twilight of the Newfoundland fishery.

On April 23, 1997, Fred Mifflin, then federal minister of fisheries and oceans, announced that there would be a "fifty-seven percent increase in the total allowable catch of northern shrimp"[2] and that the overall management plan for the next two years included a special allocation of three thousand tonnes for the Great Northern Peninsula. In a matter of weeks, St. Anthony Basin Resources Inc. (SABRI) was formed. Made up of a fifteen-member volunteer board comprising five fisherpersons, four fish plant employees, four community representatives, and two representatives from communities on the Northern Peninsula, SABRI had to quickly decide the best way of going about establishing a system to manage the special allocation, bearing in mind that "one thing had to remain true, the shrimp had to be caught in such a way that it would provide the greatest benefit to the residents of the area."[3]

SABRI placed an advertisement in newspapers province-wide, looking for both short-term (as the shrimp had to be out of the water by year's end) and long-term proposals. The latter included a stipulation that the successful partnering company had to agree to establish a processing facility for shrimp and other species in St. Anthony. After a prolonged review process, on October 31, 1997, a fifteen-year, fifty percent ownership contract was signed with 10635 NF Ltd., a partnership between GNP Fisheries and Clearwater Fine Foods. The multimillion-dollar shrimp plant began operations in 1999.

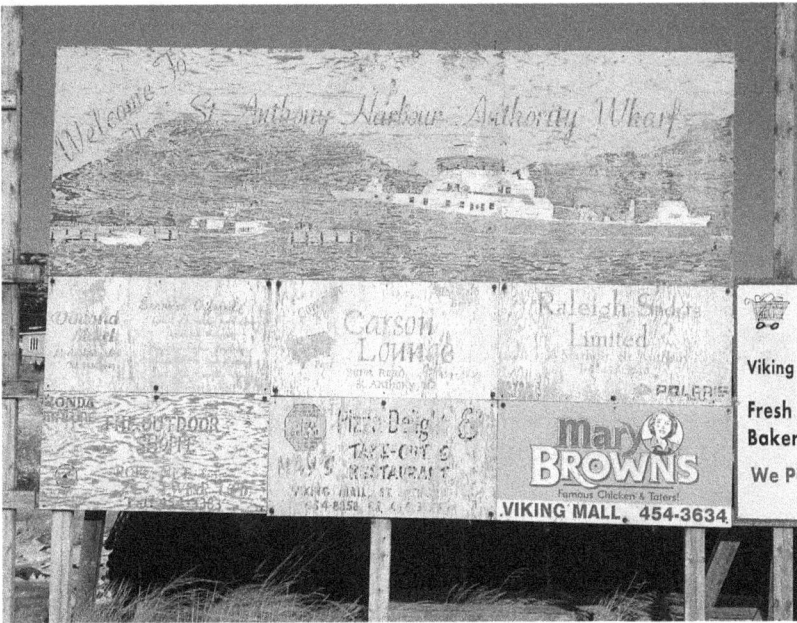

"Welcome to St. Anthony Harbour Authority Wharf," St. Anthony, Newfoundland and Labrador, 2011. Photograph by author.

While in the years preceding the cod moratorium St. Anthony had a dwindling small-boat fishery, it is now entirely gone. Draggers, owned by Clearwater, provide employment for local fisherpersons who work on the boats as crew and ensure that the shrimp-processing facility has a ready supply, as the draggers' route takes them from Newfoundland to Labrador and on to Greenland. Small-boat fishing here has now become a leisure pastime, on par with moose and duck hunting or cross-country treks on all-terrain vehicles. From April until November the processing facility employs an additional 215 people who cook more than fourteen million pounds of shrimp per season.[4] The average age of an employee at the plant is fifty-five.[5]

Ultimately, SABRI's structuring process of revenue sharing has brought a flagship investment to the region, which in turn has attracted further federal and provincial funding for other infrastructure projects; it has made St. Anthony a big fish in a diminishing pond on the Northern Peninsula. Outport communities located a mere five to ten kilometers away, such as Great Brehat and Little Brehat, are today simply monuments to the way it was. New houses are going up in St. Anthony. The last small-boat fisherperson,

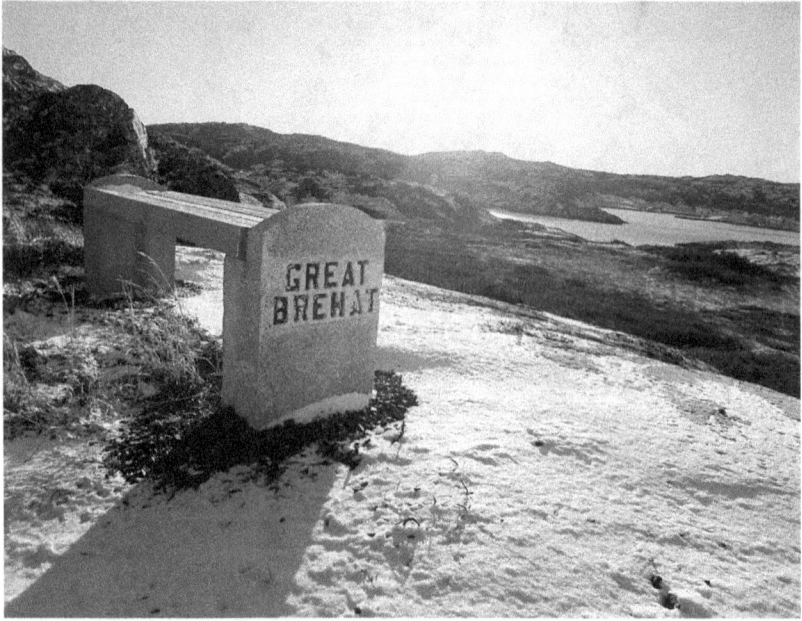

Lookout bench angled toward Great Brehat. St. Anthony, Newfoundland and Labrador, 2011. Photograph by author.

like many before, went into contracting. And yet the regional management of this finite resource has ushered in an era, a very recent era, of new road building, the planning and construction of more extensive tourist infrastructure around the Viking archeological site of L'Anse-aux-Meadows, and talk of strategies for fostering winter tourism; "this has created a different economy for us."[6]

Curtis Richards, who was town manager of St. Anthony when I interviewed him in 2011, reminded me that the SABRI revenue sharing scheme is no different from the deal demanded by many communities looking for their share in the province's oil projects. It is about "security" and "growth" for the community, he said. Negotiating the zones of engagement of a natural resource is a difficult interplay between provincial and municipal concerns "because if you're developing the resource just to obtain funds for your government to operate as it sees fit, you're doing no different than the people in Venezuela or the people in Syria or anywhere else. You're taking a resource from the people and you're using it for your own good and the people are suffering because of it. So if you develop it, you don't have those problems."[7]

St. Anthony Seafoods cold storage facility, St. Anthony, Newfoundland and Labrador, 2011. Photograph by author.

This is a live economic philosophy the likes of which Gibson-Graham both fosters and takes apart. While it is contingent on a multiplicity of scales, from that of the individual and communities to corporations and global markets, it is nonetheless a form of live economics in that its outcomes are ongoing and dependent on very definite material circumstances and decisions. Locations, for Gibson-Graham, are also important conjunctures of actors.

In an echo of its evolving historical valences that I touched on over the course of this book, the fishery in Newfoundland and Labrador today could not be closer in kind to Berlant's understanding of "cruel optimism": that "condition of maintaining an attachment to a significantly problematic object."[8] As with so many "problematic objects," the fishery is a concatenation of promises, of still-to-be-negotiated infrastructural mediations and any number of subject positions. It is a placeholder for the evolving promises of extraction: a staple economy in its unnatural evolution from fish to oil to fish. The relationship that St. Anthony, as a difficult-to-discern agential whole, has constructed with that set of promises is that of a resolute realist, its realism stemming from the type of attitude that still sees the possibility of a seal fishery coming to fruition. And yet it is at once a realism that can make out how much future possibilities depend on that "political game, and that political game is still ongoing, with the seals, and with the cod, and with the crab and the shrimp, it's all political. We are benefiting from some of the decisions that have been made with respect to the area, and a lot of it comes from the past from which we had to learn to survive. And I think that instinct for survival is in everybody you find in this region today."[9] This "never say die" attitude is a distinctly, if problematically, optimistic one. It is a reminder of the finiteness, in its double sense, of both resources *and* economies—that is, limited in both quantity and scope.

INTRODUCTION

1. Galloway, Thacker, and Wark, *Excommunication*, 2.

2. See Bolter and Grusin, *Remediation*.

3. Guillory, "Genesis of the Media Concept."

4. See Parks, *Rethinking Media Coverage*.

5. Tsing, *Friction*, 35.

6. Berlant, "Commons," 393.

7. Kember and Zylinska, *Life after New Media*, 42.

8. *Toilers of the Deep* was the official magazine of the Royal National Mission to Deep Sea Fishermen, a Protestant missionary organization that originally sponsored Grenfell to visit the colony of Newfoundland in 1892.

9. Of late the proper noun "Grenfell" has become synonymous with the Grenfell Tower fire that devastated North Kensington, London, in 2017. The tower took its name from Grenfell Road that ran along its southern edge, with the thoroughfare named after Field Marshal Francis Wallace Grenfell, a British army officer who was active in imperial military activities in North Africa in the late nineteenth century. There is no real parallel to draw between the neglect that caused the 2017 fire and the Grenfell Mission story. They both reside in the semantic resonance of this British proper noun; stories that inhere in the commemorative and everyday sites of the British imperial project.

10. Rompkey, *Grenfell of Labrador*, xiv.

11. Stoler, *Along the Archival Grain*, 7.

12. See Tsing, *Mushroom at the End of the World*.

13. Tsing, *Friction*, 37.

14. Parks and Starosielski, introduction, 5.

15. Stoler, *Along the Archival Grain*, 7.

16. For an articulation of this tension in a parallel subarctic context, see Piper, *Industrial Transformation of Subarctic Canada*.

17. Cited in Cameron, *Far Off Metal River*, 17.

18. See Bavington, *Managed Annihilation*.

19. Tsing, *Friction*, 36.

20. See Gómez-Barris, *Extractive Zone*.

21. The first Moravian settlement was established in Nain in 1771, with the pres-

ence of members of this Protestant Episcopal church in Labrador dating to the mid-eighteenth century. The Moravian sphere of influence in the region would continue for nearly three centuries and constituted the first Christian mission to an Inuit community on the territory of what is now known as Canada. See "The Moravian Church," *Newfoundland and Labrador Heritage*, http://www.heritage.nf.ca/articles/society /moravian-church.php; and Rollman, *Labrador through Moravian Eyes*.

22. See Grenfell, *Vikings of Today*.

23. Warwick Anderson's analogous treatment of American physicians working in the Philippines highlights how colonized Filipino bodies had to be set apart and differentiated from the potentials of white physiology: "In the tropics, then, American scientists and physicians felt compelled to reinvent their whiteness and harden their masculinity. Alongside the science of native pathology, health officers developed a positive and perhaps sadly overassertive science of white physiology and mentality. White male bodies and white male minds were repeatedly differentiated from those of Filipinos and insulated from apparently hostile and degenerative surroundings, especially from moist heat, germs, and Filipino social life. Physicians sought to construct a white corporeal armature—a hard, sporty indifference—to their multiply challenging milieu. But often their whiteness and manliness proved disappointingly fragile or corruptible." Anderson, *Colonial Pathologies*, 7.

24. Tsing, *Friction*, 37.

25. Walker and Starosielski, "Introduction: Sustainable Media," 17.

26. Peters, *Marvelous Clouds*, 3.

27. Peters, *Marvelous Clouds*, 14.

28. Peters, *Marvelous Clouds*, 17.

29. Starosielski, *Undersea Network*, 17–18.

30. Prowse, *History of Newfoundland*, xix.

31. "Report of the Superintendent," *First Annual Report of the International Grenfell Association* (1914), 10–11, MG 63.2207, Grenfell Mission Leaflets and Booklets, New England Grenfell Association Records.

32. The mission's industrial department was essentially a homecraft industry that sought to combine the advantages of a readily available labor force (women at home, especially in the winter months; patients convalescing; fisherfolk suffering a debilitating illness or injury that did not allow them to return to the fishery and so had to find other employment) and materials (wood and a locally available stone known as labradorite). The department, as one of the most enduring legacies of the mission (handicrafts are still a part of the local economy; see Grenfell Interpretation Centre, Grenfell Handicrafts Online Store, accessed May 27, 2020, http://www.grenfell-properties.com /handicrafts), which tells a fascinating story that connects Jessie Luther, a pioneer in the occupational health movement and an early contributor to Jane Addams's Hull House movement in Chicago, with voluntary export campaigns of silk stockings from metropolitan centers in the United States and Canada destined for northern Newfoundland and Labrador and the production of hooked mats. For a more comprehensive account, see Laverty, *Silk Stocking Mats*; and Rompkey, *Jessie Luther at the Grenfell Mission*.

33. This ambitious project was opened in 1911. Earlier that same year, at the outset of

construction, King George V, upon pushing a button in London, laid the building's cornerstone via an early relaying electrical mechanism. Made of brick and reinforced concrete, on designs donated by the prominent New York architectural firm of Delano and Aldrich, and with a substantial estimated construction cost of $150,000, it occupied a prominent site in the center of St. John's. Intended to function as affordable lodgings run at a reasonable rate, it also was meant to keep the itinerant sailors, fishermen, and female workers, among others, out of the nefarious drinking establishments and lodging houses near the waterfront. "The basement contained a cobbler's shop where men could repair their shoes, a laundry where they could wash their clothes, a forty-five foot swimming pool, and a bowling alley. The main floor featured a large hall that could seat 350, a dining-room, a reading room, and a temperance bar. The third and fourth floors consisted of bedrooms, the upper floor reserved exclusively for the use of women in the fishing industry." Rompkey, *Grenfell of Labrador*, 171. It was a white elephant of a project from the start, and problems persisted in its management.

34. See Ruiz, "Grenfell Cloth."

35. Cited in Parks and Starosielski, introduction, 14.

36. Parks and Starosielski, introduction, 14.

37. Kember and Zylinska, *Life after New Media*, xiii. Richard Grusin's understanding of "radical mediation" is in line with this focus on the processual and evolutionary qualities of mediation as a philosophical category, however he foregrounds, through Alfred North Whitehead, Gilbert Simondon, and Karen Barad, the term's "ontogenetic" capacities and how it can exceed the retention of "communication" as its paradigmatic twentieth-century reference point; see Grusin, "Radical Mediation."

38. See Gabrys, *Program Earth*.

39. Kember and Zylinska, *Life after New Media*, 21.

40. Kember and Zylinska, *Life after New Media*, 19.

41. Kember and Zylinska, *Life after New Media*, 19.

42. Kember and Zylinska, *Life after New Media*, 1.

43. Cameron, *Far Off Metal River*, 12.

44. Sobchack specifies that these "family features" include "a valorization of media in their concrete particularity rather than as a set of abstractions; media as material and structures (in their broadest and most dynamic sense) rather than as subaltern 'stuff' subject (and subjected) to theory or metaphysics; media practice and performance as a corporeal, instrumental, and epistemic method productively equal to methods of distanced analysis; description of media's materials, forms, structures, and operations rather than the interpretation of media content or social effects; media's formal and epistemic variety rather than their remedial similitudes; and, finally (at least in this litany), media, in their multiplicity, rupturing historical continuity and teleologies rather than supporting them." Sobchack, "Afterword," 327.

45. Cameron, *Far Off Metal River*, 25.

46. See Parrikka, *Medianatures*.

47. See Toland, *A Sort of Peace Corps*.

48. See Spice, "Fighting Invasive Infrastructures"; and Dafnos and Pasternak, "How Does a Settler State Secure the Circuitry of Capital?"

49. Stoler, *Duress*, 5.

50. Stoler, *Along the Archival Grain*, 1.

51. I first came across this phrase during a tour of Francis Patey's basement workshop. A local historian in St. Anthony, Patey has written a number of books on the Grenfell Mission and life in the region in the early twentieth century. The phrase appeared on cut-out panels that were glued to the front of miniature wooden fishing stages that Patey was making to sell at St. Anthony's Come Home Year event in 2013.

52. Stoler also equates consequential "minor" histories with Michel Foucault's description of the statement-event: what "emerges in its *historical irruption*; what we try to examine is *the incision* that it makes, the irreducible—and very often tiny— emergence." Stoler, *Along the Archival Grain*, 7. This event-driven character of minor history is also reminiscent of Kember and Zylinska's notion of the "cut" (following the analogous poststructuralist tradition of Jacques Derrida) that media studies scholars must make when trying to parse the media flow of mediation; see Kember and Zylinska, *Life after New Media*, xvi.

53. From the scholarly to the sensationalistic, the mission has been treated across a number of genres of study and works of the imagination. Ronald Rompkey's scholarly work is the touchstone in the field, and while I have already noted several of the memoirs he has edited, I should also include *Labrador Odyssey* and *The Labrador Memoir of Dr. Harry Paddon*. The accounts of volunteer nurses, and they are numerous, are mostly to be found in issues of the International Grenfell Association journal, *Among the Deep Sea Fishers*. A few book-length exceptions are Elliot Merrick's *Northern Nurse*, a recounting of his wife's experiences working for the mission in Labrador; Edith Tallant's work of fiction, *The Girl Who Was Marge*; and Patricia O'Brien's edited anthology, *The Grenfell Obsession*. A later account of the mission, given by one of its principal physicians, can be found in Gordon Thomas's *From Sled to Satellite*. For an overview of recent mission literature see Hiller, "Grenfell and His Successors." Finally, St. Anthony– born former forest ranger turned novelist Earl B. Pilgrim has dramatized many of Grenfell's most adventurous encounters and endeavors in *The Captain and the Girl* and *The Day Grenfell Cried*.

54. Edwards, "Infrastructure and Modernity," 186.

55. Peters, *Marvelous Clouds*, 38.

56. See, by way of example, Maxwell and Miller, *Greening the Media*; and Bozak, *Cinematic Footprint*.

57. Parks and Starosielski, introduction, 14.

58. Hayles, "Simulated Nature and Natural Simulations," 413.

59. Heise specifies: "The analysis of the virtual territories of media environments remains incomplete without a consideration of the competing material territories of built and natural environments. Just as environmentalists need to address the ways in which recent technologies have altered our experience and conceptualization of the natural, media theorists need to find ways of relating the global connectedness of virtual space back to the experiences of physical space that individuals and communities simultaneously undergo. Such a move toward a more general ecology of space would be an important step in the 'greening' of media ecology, as well as in the investigation

of 'posthuman' identities that unfold at the interface of nature and technology." Heise, "Unnatural Ecologies," 168.

60. For an analogous treatment of a media theory submerged across its aquatic dimensions, see Jue, *Wild Blue Media*.

61. See Innis, *History of the Canadian Pacific Railway*, *Fur Trade in Canada*, and *Cod Fisheries*.

62. See Star and Ruhleder, "Steps toward an Ecology of Infrastructure."

63. Young, "Innis's Infrastructure," 240.

64. See Evenden, "Northern Vision of Harold Innis"; Berland, *North of Empire*; and van Wyck, *Highway of the Atom*. As Berland specifies: "Innis theorizes place as a spatial, temporal, and economic entity but takes into account no living place in particular. The technologies that produce his centers and margins never encounter the everyday lives, the complex mediated power dynamics, the lively vestiges of myth and memory, the diverse imaginative activities of real men and women. It is this omission, not his attention to technology and space in the history and practice of empire, that makes Innis vulnerable to the charge of determinism." Berland, *North of Empire*, 96. More recently, Young nuances Innis's examination of the fur trade economy by emphasizing how "colonization was enacted as a slow, sedimentary, and violent historical event." Young, "Innis's Infrastructure, 240.

65. Peters, *Marvelous Clouds*, 18.

66. As Macarena Gómez-Barris notes: "Therefore, the extractive view sees territories as commodities, rendering land as for the taking, while also devalorizing the hidden worlds that form the nexus of human and nonhuman multiplicity. This viewpoint, similar to the colonial gaze, facilitates the re-organization of territories, populations, and plant and animal life into extractible data and natural resources for material and immaterial accumulation." Gómez-Barris, *Extractive Zone*, 5.

67. See Barney, "To Hear the Whistle Blow"; Bonnett, *Empire and Emergence*; Buxton and Acland, *Harold Innis in the New Century*; Buxton, *Harold Innis and the North*; and van Wyck, *Highway of the Atom*.

68. Innis, *Cod Fisheries*, 482.

69. As Innis notes, systems derived from commercialism were often heavily subsidized, relied on protectionist measures, and favored short-term credit. By way of contrast, the touchstones of free market competition, a reliance on the acquisition of technologies to increase efficiencies in production and distribution and long-term credit were the defining characteristics of capitalism that were gradually integrated into the colonial fishery. Innis, *Cod Fisheries*.

70. Innis, *Cod Fisheries*, 482.

71. Berland, *North of Empire*, 78.

72. Tsing, *Friction*, 37

73. Tsing, *Friction*, 37

74. See Barney, "To Hear the Whistle Blow"; and Starosielski, *Undersea Network*. Barney calls for the emergence of a form of "critical *agricultural* studies." Barney, "To Hear the Whistle Blow," 7.

75. Van Wyck, *Highway of the Atom*, 19.

76. Cited in Rompkey, *Grenfell of Labrador*, 38.

77. Cited in Rompkey, *Grenfell of Labrador*, 41.

78. Letter from Wilfred Grenfell to E. A. B. Willmer, February, 23, 1931, MG 63.164, Grenfell Letters, Grenfell Association of Great Britain and Ireland Papers.

79. This dimension of the Grenfell Mission story echoes Edward Jones-Imhotep's recent articulation of the post–World War II Canadian nation-state as both a "natural object defined by distinctly 'Northern' upper-atmospheric phenomena" as well as "a technological space of uniquely powerful, widespread radio failures that threatened the technological integrity of the nation." The mission sought to respond to its surrounding ecological conditions while also trying to improve the extractive relationships fisherfolk were fashioning in response to them. "Decay, degradation, wear, cracking, and corrosion are also 'natural' processes," Jones-Imhotep writes. "Rather than opposing nature or blending seamlessly with it, technologies have figured throughout history as media whose problematic behavior expresses and even defines the natural. If technology is society made durable, then there are crucial episodes, signal instances, where nature is technology made fallible." Jones-Imhotep, *Unreliable Nation*, 3–4 and 12–13.

80. Gregory Dreicer traces the relationships between infrastructure and nation building: "Technological objects serve as ideal containers for nationalistic views. They allow feelings about nativenesss and foreignness to assume a tangible form. Moreover, infrastructure does seem to reflect the state of the nation by demonstrating a government's ability to maintain the networks that enable the nation to function." Dreicer, "Building Bridges and Boundaries," 162.

81. Peters, *Marvelous Clouds*, 37.

82. Cowen, "Infrastructures of Empire and Resistance."

83. Cowen, "Infrastructures of Empire and Resistance." Andrew Barry picks up on a similar thread spanning the posts that appear in *Cultural Anthropology*'s the Infrastructure Toolbox series: "The infrastructures glimpsed here have a history or, rather, they are the products of multiple histories." Barry, "Discussion: Infrastructural Times."

84. Ashley Carse notes that geographer Andrew Barry "uses the term *infrastructural zones* to describe the emergent and striated spatial forms created by the reduction of differences between systems via common connection standards." Carse, *Beyond the Big Ditch*, 12.

85. Cowen, "Infrastructures of Empire and Resistance."

86. Carse writes: "I emphasize infrastructural work—a term used by science, technology, and society scholar Geoffrey Bowker—to foreground the variety of organizational techniques (technical, governmental, administrative, environmental) that create the conditions of possibility for rapid and cheap communication and exchange across distance. As scholars have shown, infrastructure does not refer to any specific class of artifact, but to a process of relationship building and maintenance." Carse, *Beyond the Big Ditch*, 11. Carse's efforts at extending Star and Ruhleder's foundational work toward its ecological dimensions highlights how infrastructure and environment making are often one and the same process.

87. "First, infrastructure is not a specific class of artifact or system, but an ongoing

process of relationship building," Carse writes. "Seen in this way, engineered canals and highways are surprisingly social and ecological. As temporary lines across active environments that erode, rust, and fracture them, infrastructures advance and retreat in relation to the capital and labor channeled into their construction and maintenance." Carse, *Beyond the Big Ditch*, 5.

88. Cited in Barney, "To Hear the Whistle Blow," 6.

89. Berlant, "Commons," 393.

90. Berlant, "Commons," 403.

91. P. J. Roberts, "Process of Change," 1.

92. P. J. Roberts, "Process of Change," 1.

93. P. J. Roberts, "Process of Change," 2.

94. See the Processing Facilities page of the Clearwater Seafoods website, accessed February 18, 2014, https://www.clearwater.ca/en/ocean-to-plate/processing-facilities.

95. Berlant nuances this understanding of the evolving relationalities between affect and the temporal horizons of its infrastructural emergence: "The political and epistemic problem for the politically autopoetic—which is what all world-creating subjects in common struggle are—is that the placeholders for our desire become factishes, fetishized figural calcifications that we can cling onto and start drawing lines in the sand with. . . . What remains for our pedagogy of unlearning is to build affective infrastructures that admit the work of desire as the work of an aspirational ambivalence. What remains is the potential we have to common infrastructures that absorb the blows of our aggressive need for the world to accommodate us and our resistance to adaptation and that, at the same time, hold out the prospect of a world worth attaching to that's something other than an old hope's bitter echo. A failed episode is not evidence that the project was in error. By definition, the common forms of life are always going through a phase, as infrastructures will." Berlant, "Commons," 414.

96. Cowen, "Infrastructures of Empire and Resistance."

THE WAY IT WAS, ST. ANTHONY, 1959

1. Patey's numerous books include *The Jolly Poker*, *The Grenfell Dock*, and *A Battle Lost*, all of which document the history of the region, its struggle to maintain a seal fishery, and many of its economic and cultural traditions.

2. Tsing, *Friction*, 28–29, emphasis added.

3. See Starosielski, *Undersea Network*, 14.

4. Susan Leigh Star and Karen Ruhleder, quoted in Parks and Starosielski, introduction, 9.

5. Berlant, "Commons," 403. Van Wyck remarks how this territorial archive is bound up with the memory work enacted by the Dene, who were employed to mine and transport uranium extracted from the Eldorado mine in the Northwest Territories and who decades later would travel to Japan to apologize for what they saw as their deferred moral complicity in the effects of the atomic bombs; dispersal occurs across the land and is registered across generations. Van Wyck, *Highway of the Atom*, 18–21.

6. See Mitchell, *What Do Pictures Want*? Mitchell emphasizes the "worldmaking" ca-

pacities of "pictures," that is, the entire apparatus that brings image, object, and medium together, for, as Mitchell claims, "there is no getting beyond pictures" and attaining a privileged relationship with "the Real" (Mitchell, *What Do Pictures Want?*, xiv). He specifies that we should turn our attention to the "lives of images" in order to examine what they "*want*"—"what claim they make upon us, and how we are to respond" (xv). It is partly in this spirit that I assemble this array of materials in *Slow Disturbance*. Yet I also want to modulate Mitchell's emphasis on a human "us" in the claims pictures make and leave the reproductions that appear here as precisely that—reproductions of places, documents, and more that retain their own nonhuman ontology that exceeds my own vision as authorial photographer. Joanna Zylinska touches on a number of these issues in *Nonhuman Photography*, particularly with regard to foregrounding photography's "ontological force and its significance as a life-shaping medium." See Zylinska, "Introduction: Capturing the End of the World," https://www.nonhuman .photography/introduction.

7. For a comprehensive summary of this sensorial turn in infrastructure studies, see Young, "Innis's Infrastructure," 243; Mattern, "Infrastructural Tourism"; and Mattern, *Deep Mapping the Media City*.

8. See Tsing, *Mushroom at the End of the World*, 160–61.

9. Stoler, *Along the Archival Grain*, 1.

10. Gitelman, *Paper Knowledge*, 4.

11. Esther Leslie cited in Gabrys, *Digital Rubbish*, 28.

12. Nixon, *Slow Violence*, 2.

13. Berlant, "Commons," 393.

14. Nixon, *Slow Violence*, 3.

15. Nixon, *Slow Violence*, 8.

16. As Gitelman notes, "If all documents share a certain 'horizon of expectation,' then, the name of that horizon is accountability." Gitelman, *Paper Knowledge*, 2.

17. Goose Bay really came into settler being at the start of World War II. At the time it was the longest airfield in the world and a massive military base that made up North America's first line of defense. This was of course a distant and abrupt reality for the fishermen working off the coast of Labrador over the early decades of the twentieth century. Supplying cod to the fish-eating countries of the world, from Portugal to Brazil to North Africa, it is an industry that has today shifted to the processing of such shellfish as snowcrab and such crustaceans as shrimp. Goose Bay on a bleak November day is a collection of utilitarian strip malls, nondescript, low-rise office buildings, and generic tract housing that is trying to keep up with demand. Here it is most definitely a resource reality that reigns. It is to the distant mines and hydroelectric projects that the city looks and travels. The week I arrived in Goose Bay, the Inuit of Nunatsiavut, a territory on the North Coast of Labrador and within the province of Newfoundland and Labrador, had just signed what was called the "New Dawn" agreement. Heralded as an equitable exploitation of the hydroelectric power produced by the Lower Muskrat Falls, it is a new form of negotiation around what could be thought of as an *indigenous resource*, with that term containing the ambiguous place that Indigenous human actors hold in the composition of dispossessed natural resources.

CHAPTER 1. THE PLANT

1. Much of the mission's built infrastructure came under the banner of "practical therapy": "A tannery, co-operative store and a haul-up slip on which schooners and small steamers are under repair all through the open water season are further instances of practical therapy, at once preventive and curative." Rompkey, *Labrador Memoir of Dr. Harry Paddon*, 17. Jane Addams's settlement house reform movement in Chicago influenced the mission's early efforts at practical therapy. See Addams, *Twenty Years at Hull House*; and Carson, *Settlement Folk*.

2. *Oxford English Dictionary*, s.v. "plant"; accessed July 21, 2020, https://www-oed-com.login.ezproxy.library.ualberta.ca/view/Entry/145156?rskey=6Qx02z&result=4&isAdvanced=false#eid.

3. Rompkey, *Labrador Memoir of Dr. Harry Paddon*, 31.

4. "Report of the Superintendent," *First Annual Report of the International Grenfell Association* (1914), 4, MG 63.2207, Grenfell Mission Leaflets and Booklets, New England Grenfell Association Records.

5. "And the Lord God planted a garden eastward in Eden; and there he put the man whom he had formed." Genesis 2:8. Here and in subsequent biblical citations, I am using the King James Version.

6. "Moreover I will appoint a place for my people Israel, and will plant them, that they may dwell in a place of their own, and move no more; neither shall the children of wickedness afflict them any more, as beforetime." 2 Samuel 7:10.

7. "And he shall be like a tree planted by the rivers of water, that bringeth forth his fruit in his season; his leaf also shall not wither; and whatsoever he doeth shall prosper." Psalm 1:3.

8. "And I have put my words in thy mouth, and I have covered thee in the shadow of mine hand, that I may plant the heavens, and lay the foundations of the earth, and say unto Zion, Thou art my people." Isaiah 51:16.

9. "Thus shalt thou say unto him, The LORD saith thus; Behold, that which I have built will I break down, and that which I have planted I will pluck up, even this whole land." Jeremiah 45:4.

10. Jackson, "Rethinking Repair," 233–36.

11. On the relationship between Protestant sermons and everyday action from the nineteenth century forward, see Jackson, *The Word and Its Witness*.

12. Cosby, "Impressions of the Mission Stations."

13. Cosby, "Impressions of the Mission Stations," 75.

14. Cosby, "Impressions of the Mission Stations," 76.

15. Cosby, "Impressions of the Mission Stations," 77.

16. Cosby, "Impressions of the Mission Stations," 76.

17. For a more detailed account of McNeil's instrumental role in completing the 1927 mission hospital in St. Anthony, see Ruiz, "Behind the Scenes at the Grenfell Mission."

18. Curtis, "St. Anthony."

19. Peters, *Marvelous Clouds*, 38.

20. Letter from Wilfred Grenfell to Sir John Hope Simpson (Colonial and Domin-

ions Office), February, 14, 1935, MG 63.164, Grenfell Letters, Grenfell Association of Great Britain and Ireland Papers

21. Jackson, "Rethinking Repair," 227.

22. For the context of Bowker's early deployment of the phrase "infrastructural inversion," see his *Science on the Run*, 10. See also Appel, Anand, and Gupta, "Temporality, Politics, and the Promise of Infrastructure."

23. See Rompkey, *Grenfell of Labrador*, 43, 57, 79, 98–99, 117–19, 164, 216, and 224.

24. Letter from Horace McNeil to Kathleen Young (IGA), May 28, 1952, MG 63.1949, Insurance files (New York), Grenfell Association of America, International Grenfell Association Fonds.

25. Cosby, "Impressions of the Mission Stations," 75.

26. Thomas, "Farewell and Godspeed," 14.

27. For an account of practices of self-medication among outport fisherfolk, see Crellin, *Home Medicine*.

28. Grenfell, *Labrador Doctor*, 120–21.

29. See Rompkey, *Grenfell of Labrador*, 53–55; Sider, *Culture and Class in Anthropology and History*; and Hiller and Neary, *Newfoundland in the Nineteenth and Twentieth Centuries*.

30. Rompkey, *Grenfell of Labrador*, 54.

31. Rompkey, *Grenfell of Labrador*, 55.

32. Cited in Rompkey, *Grenfell of Labrador*, 53.

33. For an overview of how this conjuncture has developed in the Canadian context see Brown, *Struggling for Effectiveness*.

34. Carse, *Beyond the Big Ditch*, 11.

35. See Innis, *Cod Fisheries*, 482.

36. Berlant, "Commons," 393.

37. Grenfell, *Labrador Doctor*, 126.

38. See Rompkey, introduction to *Labrador Odyssey*, xxvi.

39. Rompkey, introduction to *Labrador Odyssey*, xx.

40. Rompkey, *Grenfell of Labrador*, 57.

41. See Mackinnon, *Vernacular Architecture in the Codroy Valley*; and Mellin, *Tilting*. The Royal National Mission to Deep Sea Fishermen is the same organization.

42. Sean O'Dea, email message to author, November 10, 2008.

43. Rompkey, *Labrador Odyssey*, 33.

44. Berlant, "Commons," 393.

45. "Biography Notes—Sir Wilfred Grenfell [Before 1957]," 23, MG 63.311, Grenfell Association of Great Britain and Ireland Papers.

46. For a discussion of how a comparable early twentieth-century medical response to such an epidemic as tuberculosis could go from being transitionally to permanently institutionalized see Connolly, *Saving Sickly Children*; and Hobday, "Sunlight Therapy and Solar Architecture."

47. "Biography Notes—Sir Wilfred Grenfell [Before 1957]," 23.

48. See Grenfell, "Lure of the Labrador."

49. Stephen Graham and Simon Marvin cited in Jackson, "Rethinking Repair," 228.

50. "Equipment on Coast—Data 1931–1948," MG 63.1978, Grenfell Association of Great Britain and Ireland Papers. For a nineteenth-century equivalent, see Briggs, *Victorian Things*.

51. For an analysis of a similar phenomenon, though in a different regional context, see McKay, *Quest of the Folk*.

52. Berlant, "Commons," 393.

53. Berlant, "Commons," 403.

54. Rompkey, introduction to *Labrador Odyssey*, xx.

55. Rompkey, *Labrador Odyssey*, 156.

56. Rompkey, *Labrador Odyssey*, 19.

57. Rompkey, *Labrador Odyssey*, 175.

58. "Biography Notes—Sir Wilfred Grenfell [Before 1957]," 23.

59. Peters, *Marvelous Clouds*, 102.

60. For a different regional context, see Adams, *Architecture in the Family Way*.

61. Rompkey, *Labrador Memoir of Dr. Harry Paddon*, 31.

62. Rompkey, *Labrador Memoir of Dr. Harry Paddon*, 31.

63. Paddon, "Indian Harbour Hospital."

64. For an account of the broader social gospel movement of which these values were an integral part, see Hopkins, *Rise of the Social Gospel in American Protestantism*.

65. Berlant, "Commons," 394.

66. Grenfell, "Doctor Grenfell's Log," 11.

67. Grenfell, "Who Will Provide for These Special Needs?" 54.

68. "In Grenfell's Labrador," *New York Times*, November 7, 1930, MS 254, box 15, Wilfred Thomason Grenfell Papers, Yale University Archives.

69. "A Brick for Labrador," 5, MG 63.2207, Grenfell Mission Leaflets and Booklets, New England Grenfell Association Records.

70. "A Brick for Labrador," 20.

71. "A Brick for Labrador," 18.

72. Peters, *Marvelous Clouds*, 38

73. Anne Spice points to current divergent understandings of "critical infrastructure" in the Canadian context, from capital-intensive and extractive projects led by the federal government to Indigenous communities' reliance on clean water, unpolluted riverways, and broader land-based ecologies. See Spice, "Fighting Invasive Infrastructures," 41.

74. Children's Page, "Children's Home, St. Anthony."

75. "New St. Anthony Hospital," MG 63.2207, Grenfell Mission Leaflets and Booklets, New England Grenfell Association Records.

76. Berlant, "Commons," 403.

77. Grenfell, *Down to the Sea*.

78. Rompkey, *Grenfell of Labrador*, 117. Grenfell would lecture to the middle and upper classes of the northeast coast, presenting via the relatively novel means of the stereoticon tinted scenes of "life on the Labrador": the poverty of its residents (good Anglo-Saxons and so readily available for the possibility of empathy—a sentiment new to the young century), the sublimity of its landscape (described by the *New York*

Times as America's Norway with a touch of Switzerland in the backdrop; see "Labrador Medical Mission," *New York Times*, April 26, 1903), and the progress of the plant. Grenfell's excursions all came in the thick of the Progressive Era in American life. To schematize its aims and its means, from the 1890s into the 1920s social reformers of all political and religious persuasions attempted to reshape the collective life of their fellow citizens through a shared belief in science, expertise, the mobilization of technology, and the inherent goods and possibilities of education. See McGeer, *A Fierce Discontent*. Later in the mission and Grenfell's life, indissociable as they were, it was the waning pace of his lecture tours that spurred the IGA administration to think about restructuring the mission as a whole. As a tally of Grenfell's lecture tour in the fall of 1928 shows, it was a lucrative if exhausting undertaking. From September 23 to December 14, Grenfell delivered seventy three lectures, going from Croydon, near London, to Endinburgh and back to Westminster, amassing through fees, collections, and sales of merchandise £3,582.19. It was a necessary form of seasonal work of Grenfell's own, as it kept the mission's work in the (donating) public's eye and also solidified Grenfell's place as an extraterritorial figure who, through his public speaking as well as book and article publishing, could inhabit both worlds, having heeded the "lure of the Labrador" while also bringing back true facts of need having been vanquished through true deeds of daring and venture. "Lecture Tour 1928," MG 63.164, Grenfell Letters, Grenfell Association of Great Britain and Ireland Papers.

79. See Naylor, *Canadian Health Care and the State*; and Hoffman, *Health Care for Some*.

80. See Grenfell, *Vikings of Today*.

81. Hornsby and Schmidt, *Modern Hospital*.

82. Hornsby, "Trend of Modern Hospital Service," 98.

83. "For the first time we have borrowed money to enable us to put a building through. Towards the $80,000 necessary, we have given the promise of $20,000." Grenfell, "New Hospital for St. Anthony," 51.

84. See Rudofsky, *Architecture without Architects*; and, by way of contrast, Davis, *Culture of Building*.

85. Grenfell, "New Hospital for St. Anthony," 52.

86. Rompkey, *Grenfell of Labrador*, 299.

87. Berlant, "Commons," 403.

88. Berlant, "Commons," 403.

CHAPTER 2. CREDIT AND COMMON SENSE

1. Peters, *Marvelous Clouds*, 21.

2. See Drache, "Introduction: Celebrating Innis," xxxii.

3. See Ommer, "Merchant Credit and the Informal Economy"; Ommer, "Truck System in Gaspé"; and Cadigan, *Hope and Deception in Conception Bay*.

4. See Redfern, *Story of the C.W.S.*, iii.

5. Lazzarato, *Making of the Indebted Man*, 61. For a broader history of debt, see Graeber, *Debt*.

6. Newell, "Credit and Common Sense," 82–83.

7. *Daily News* article, quoted in Newell, "Credit and Common Sense," 83.

8. "The Social Gospel," Rompkey writes, "constituted a reforming element in late nineteenth and early twentieth century religious thinking that brought scientific knowledge and historical criticism to bear on theological ideas. Its aim was to bring Christian energy to bear upon the social problems of the time, especially those associated with urban living and the conditions of labour. The leaders were largely evangelicals or Christocentric liberals whose approach to social issues was shaped politically within the Progressive movement. And though the more radical espoused some form of socialism, in general they were progressive and reformist rather than revolutionary, aiming their churches in the direction of what Grenfell called 'public service.'" Rompkey, *Grenfell of Labrador*, 192–93; see Handy, *Social Gospel in America* and *Christian America*. In the Canadian context, a figure such as J. S. Woodsworth embodies a commitment to the social gospel movement. Woodsworth, a former Methodist preacher, first party president of the Co-operative Commonwealth Federation, was also a longtime firsthand actor in the trenches of newly forming areas, such as 1930s Winnipeg, of urban poverty. He also understood and promoted the role of the social gospel in fostering labor movement solidarities. In his sermon the "Sin of Indifference," he castigates his listeners, telling them that the greater wealth amassed, the less the sense of responsibility. Grenfell's own investments in the social gospel movement in America and Britain are, a continent apart, echoed by Woodsworth: "It is quite right for me to be anxious to save my never dying soul; but it is of greater importance to try to serve the present age. Indeed, my friend, you will save your own precious soul only as you give your life in the service of others. . . . If it is right to help the sick it is right to do away with filth and overcrowding and to provide sunlight and good air and good food. We have tried to provide *for the poor*. Yet, have we tried to alter the social conditions that lead to poverty? . . . You can't separate a man from his surroundings and deal separately with each. What nonsense we talk!" See McNaught, *Prophet in Politics*, 25–26.

9. Grenfell's log, quoted in Newell, "Credit and Common Sense," 83.

10. William Coaker founded the FPU in 1908, with the movement eventually expanding into colonial party politics in order to represent the interests of fishermen at Government House. "It was called the 'Fisherman's Union,'" as Grenfell writes, "and its apostle and mentor was a working man, who at first had no connection with politics, and whose remarkable devotion and ability led to a really notable development. Amongst other things he opened a series of union stores on the chain system that is growing so in vogue in America." Grenfell, "Labrador's Fight for Economic Freedom," 10. See Macdonald, *"To Each His Own"*.

11. Ommer, introduction to *Merchant Credit and Labour Strategies*, 9.

12. George W. Hilton, quoted in Ommer, introduction to *Merchant Credit and Labour Strategies*, 13. See Hilton, *Truck System*.

13. Ommer, introduction to *Merchant Credit and Labour Strategies*, 13.

14. Grenfell, *Labrador Doctor*, 412.

15. Grenfell, *Labrador Doctor*, 411.

16. Grenfell, *Labrador Doctor*, 412.

17. Peters, *Marvelous Clouds*, 17.

18. Peters, *Marvelous Clouds*, 14.

19. See Sterne, "Transportation and Communication."

20. Peters, *Marvelous Clouds*, 3.

21. Grenfell, *Labrador Doctor*, 423.

22. Letter from Wilfred Grenfell to Cecil Ashdown, January 13, 1938, MG 63.164, Grenfell Letters, Grenfell Association of Great Britain and Ireland Papers.

23. Letter from Wilfred Grenfell to Cecil Ashdown, January 13, 1938.

24. George Creel, "The Insurgent Sunday-School," *Everybody's Magazine* (n.d.), 471–82, MS 254, Wilfred Thomason Grenfell Papers.

25. Grenfell, *Labrador Doctor*, 422.

26. Grenfell, *Labrador Doctor*, 218.

27. Grenfell, *Labrador Doctor*, 218.

28. Letter from Francis Hopwood to Editor of *Toilers of the Deep* (7:34–44), MG 63.307, Biography Notes—Wilfred Grenfell, Grenfell Association of Great Britain and Ireland Papers.

29. Letter from Francis Hopwood to Editor of *Toilers of the Deep*, 3.

30. Letter from Francis Hopwood to Editor of *Toilers of the* Deep, 13.

31. Letter from Francis Hopwood to Editor of *Toilers of the Deep*, 13.

32. Letter from Francis Hopwood to Editor of *Toilers of the Deep*, 1.

33. Grenfell, "Labrador's Fight for Economic Freedom," 3.

34. See Roberts, "Darwinism, American Protestant Thinkers, and the Puzzle of Motivation."

35. Grenfell, "Labrador's Fight for Economic Freedom," 3.

36. See Spencer, *Principles of Biology*.

37. See Ray, *Canadian Fur Trade in the Industrial Age*.

38. Ray, *Canadian Fur Trade in the Industrial Age*, 13.

39. Ray, *Canadian Fur Trade in the Industrial Age*, 12.

40. "Co-operative Stores among Atlantic Fishermen," 9.

41. "Co-operative Stores among Atlantic Fishermen," 14.

42. Grenfell, *Labrador Doctor*, 219.

43. Grenfell, "Labrador's Fight for Economic Freedom," 21–22.

44. Grenfell, "Labrador's Fight for Economic Freedom," 33.

45. Grenfell, *Labrador Doctor*, 219.

46. Grenfell, *Labrador Doctor*, 218.

47. Grenfell, "Labrador's Fight for Economic Freedom," 14–15.

48. "The cod fish is the current coin of the fishermen. They understand its value for the purposes of exchange; but it is curious that the men do not appreciate the value of any other kind of fish (except 'bait' fish), and refuse or neglect to catch anything but cod. It is impossible to obtain in the Colony, as a rule, any fish but cod for the table. Soles, turbot, plaice, and halibut are caught, but the fishermen throw these fish back into the sea because they are not cod, and do not represent, to their minds, the market value of that fish." Letter from Francis Hopwood to Editor of *Toilers of the Deep*, 4.

49. Hiller, "Newfoundland Credit System," 98.

50. Lazzarato, *Making of the Indebted Man*, 11.

51. "Petition Presented to House of Assembly, July 12th, 1917: Re Certain Alleged Operations of the International Grenfell Association," MG 63.1947, Evidence Taken on the Grenfell Inquiry Minutes, Grenfell Association of Great Britain and Ireland Papers.

52. "Petition Presented to House of Assembly, July 12th, 1917."

53. "Counter-Petition Presented to House of Assembly by the International Grenfell Association, August 16th, 1917," MG 63.1947, Evidence Taken on the Grenfell Inquiry Minutes, Grenfell Association of Great Britain and Ireland Papers.

54. "Counter-Petition Presented to House of Assembly by the International Grenfell Association, August 16th, 1917."

55. Grenfell, *Labrador Doctor*, 224.

56. Letter from Wilfred Grenfell to Cecil Ashdown, January 13, 1938.

57. Letter from Wilfred Grenfell to Cecil Ashdown, January 13, 1938.

58. "Proposed Revision of Memorandum and Articles of Association of International Grenfell Association," MG 63.1970, RNMDSF 1916–35, Transfer of Title (New York Permanent files), International Grenfell Association Fonds.

59. "5. To acquire by purchase, concession, exchange, or other legal method, and to construct, erect, operate and maintain and manage and to sell, assign, transfer or otherwise dispose of all factories, mills, warehouses, depots, machine shops, engine houses, docks, wharves, dry docks, graving docks, railways, tramways, sidings, loading and unloading equipment, bridges, dams, reservoirs, canals, watercourses and other structures and erections, necessary for its activities or capable of being conveniently and profitably operated in connection therewith and all other property real and personal necessary or useful for the purposes of the association, and to lease, sell or dispose of the same." "Proposed Revision of Memorandum and Articles of Association of International Grenfell Association."

60. "Proposed Revision of Memorandum and Articles of Association of International Grenfell Association."

61. "(11) To draw, make, accept, endorse, execute, and issue promissory notes, bills of exchange, bills of lending, warrants, warehouse receipts, and other negotiable or transferable instruments.

"(12) To borrow and raise money in such manner as the Association may think fit.

"(13) To purchase, lease, or otherwise acquire the whole or any part of the business, property, franchises, good-will, rights or privileges held or enjoyed by any person or firm or by any corporation carrying on any business which the Association is authorized to carry on or possessed of property suitable for the purpose of this Association in such manner as may from time to time be determined.

"(14) To raise and assist in raising money for and to aid by way of bonus, loan, promise, endorsement, guarantee of bonds, debeatures [*sic*] of other securities, shares or otherwise, any other company or corporation with which the Association may have business relations and to guarantee the performance of contracts, by any such corporation or company or any other person or persons with whom the Association may have business relations.

"(15) To effect insurance against risk or loss to the Association from any cause and to insure any servants of the Association against risk or accident in the course of their employment and to insure the lives of such servants and to effect such insurance by contracts of inter-insurance or otherwise.

"(16) To invest any moneys of the Association not immediately required for the purposes of its objects, in such manner as may time to time be determined." "Proposed Revision of Memorandum and Articles of Association of International Grenfell Association."

62. "Colonial development out of a staple trade, according to a variety of scholars," Ommer writes, "rests on the creation of a set of linkages which will ultimately generate an aggregated multiplier mechanism from which the 'take-off into self-sustained growth' can develop. . . . Isolation may have been instrumental in the maintenance of the truck system in Gaspé, but the bulk of the responsibility for suppression of final demand linkages (through the reduction of real earnings and hence purchasing power) must lie with the system itself and with the conservative entrepreneurial vision of fish-merchant capital operating in a colonial context." Ommer, "Truck System in Gaspé," 71–72.

63. Grenfell, *Labrador Doctor*, 225.

64. Grenfell, *Labrador Doctor*, 225.

65. Lazzarato, *Making of the Indebted Man*, 31.

66. Peters, *Marvelous Clouds*, 177.

67. Redfern, *Story of the C.W.S.*, 313.

68. Redfern, *Story of the C.W.S.*, 313.

69. Redfern, *Story of the C.W.S.*, 313.

70. "Second factor of importance: household production; non-commercial counterpart to the staple economy; vital and pragmatic role, for both merchant and labour, played by the subsistence economy in the opening up of new lands, the development of maturing economies, and the survival of peripheral areas alike." Ommer, introduction to *Merchant Credit and Labour Strategies*, 14–15.

71. Redfern, *Story of the C.W.S.*, 357.

72. See Ryan, *Ice Hunters*.

73. Grenfell, *Forty Years for Labrador*, 145.

74. Grenfell, *Forty Years for Labrador*, 145.

75. Grenfell, "Labrador's Fight for Economic Freedom," 25–27.

76. Grenfell, "Labrador's Fight for Economic Freedom," 31.

77. Grenfell, *Forty Years for Labrador*, 145.

78. Innis, *Cod Fisheries*, 462–463.

79. Innis, *Cod Fisheries*, 494.

80. Innis, *Cod Fisheries*, 482.

81. Innis, *Cod Fisheries*, 482. Innis devotes chapter 14 to "Capitalism in Newfoundland, 1886–1936." In what is perhaps Innis's most political stance in *Cod Fisheries*, the chapter opens with the following epigraph: "'To have abandoned the principle of democracy without accomplishing economic rehabilitation is surely the unforgivable sin.' Thomas Lodge, Dictatorship in Newfoundland." Innis, *Cod Fisheries*, 449.

82. Innis, *Cod Fisheries*, xiv.

83. Hiller, "Newfoundland Credit System," 90.

84. Grenfell, "Labrador's Fight for Economic Freedom," 3.

85. Letter from Wilfred Grenfell to Cecil Ashdown, January 13, 1938.

86. Deleuze, quoted in Lazzarato, *Making of the Indebted Man*, 88.

87. Grenfell Association of Great Britain and Ireland, "Medical Work in Northern Newfoundland and Labrador: Eighth Annual Report, 1933–1934," 11, Memorial University Centre for Newfoundland Studies, Digital Archives Initiative, accessed February 6, 2013, http://collections.mun.ca/cdm4/document.php?CISOROOT=/moravian &CISOPTR=18348&REC=5.

88. Grenfell, "Sir Wilfred's Log: Summer of 1934," 99.

89. Smallwood and Pitt, "Margaret Digby."

90. Letter from Wilfred Grenfell to the Editor of the *Observer's Weekly*, St. John's, November 1937, 4, MG 63.164, Grenfell Letters, Grenfell Association of Great Britain and Ireland Papers.

91. Letter from Wilfred Grenfell to R. A. Palmer, February 20, 1934, MG 63.164, Grenfell Letters, Grenfell Association of Great Britain and Ireland Papers.

92. Lazzarato, *Making of the Indebted Man*, 21.

93. Letter from William Y. Pike to Wilfred Grenfell, March 26, 1938, MG 63.164, Grenfell Letters, Grenfell Association of Great Britain and Ireland Papers.

94. Letter from William Y. Pike to Wilfred Grenfell, March 26, 1938.

CHAPTER 3. META INCOGNITA

1. Note of "February 27, 1931," box 97, folder 2151, Correspondence Re: Sponsorship, 1927–1932, Alexander Forbes Papers.

2. Note of "February 27, 1931."

3. Note of "February 27, 1931."

4. Forbes et al., *Northernmost Labrador Mapped from the Air*.

5. Letter from Alexander Forbes to Richard Squires, box 97, folder 2157, Labrador Expedition, 1931 Correspondence with Canadian Officials, Jan 1930–Nov 1932, Alexander Forbes Papers.

6. Chow, *Age of the World Target*, 34.

7. "Fairchild Aerial Camera Corporation," box 97, folder 2152, Labrador Expedition, 1931, Correspondence and Notes Re: boat selection, Alexander Forbes Papers.

8. "Fairchild Aerial Camera Corporation."

9. Carse, *Beyond the Big Ditch*, 5.

10. See Kurgan, *Close Up at a Distance*.

11. Parks and Starosielski, introduction, 5.

12. Forbes et al., *Northernmost Labrador Mapped from the Air*, 15.

13. Forbes et al., *Northernmost Labrador Mapped from the Air*, 26.

14. Draft of the opening paragraph of *Northernmost Labrador Mapped from the Air*, 1936, box 44, folder 1536, Re: "Northernmost Labrador Mapped from the Air," 1938 (no. 1 of 23 folders), Alexander Forbes Papers.

15. Letter from Raye P. Platt to Alexander Forbes, February 1, 1937, box 44, folder

1539, Re: "Northernmost Labrador Mapped from the Air," 1938 (no. 4 of 23 folders), Alexander Forbes Papers.

16. Letter from Raye P. Platt to Alexander Forbes, February 1, 1937.

17. Forbes et al., *Northernmost Labrador Mapped from the Air*, 1.

18. Forbes et al., *Northernmost Labrador Mapped from the Air*, xv.

19. Forbes et al., *Northernmost Labrador Mapped from the Air*, 1.

20. Parks specifies the set of relationships that the drone mediates into being: "As it hovers above the earth, it can change movements on the ground. As it projects announcements through loudspeakers, it can affect thought and behavior. And as it shoots hellfire missiles, it can turn homes into holes and the living into the dead. Irreducible to the screen's visual display, the drone's mediating work happens extensively and dynamically through the vertical field—through a vast expanse that extends from the earth's surface, including the geological layers below and built environments above, through the domains of the spectrum and the air to the outer limits of orbit. The point here is that drones do not simply float above the surface of the earth—they rewrite and reform life on earth in a most material way." Parks, "Drones, Vertical Mediation, and the Targeted Class," 232.

21. See Cahill, "Brock Process of Aerial Mapping."

22. See Saint-Amour, "Applied Modernism."

23. Forbes et al., *Northernmost Labrador Mapped from the Air*, 24.

24. Forbes et al., *Northernmost Labrador Mapped from the Air*, 58.

25. Forbes et al., *Northernmost Labrador Mapped from the Air*, 59.

26. Miller, "Planetabling from the Air," 202.

27. Miller, "Planetabling from the Air," 165.

28. As Armand Mattelart notes in relation to the development of the British Empire's global communications networks, particularly over the course of the nineteenth century, "The relations of domination between centre and periphery would be etched into the very networks of national communication within dependent zones—extroversion and outside-oriented configurations would be the rule; the necessity of establishing liaisons between ports, mines, and other deposits of raw materials accounts for the rests of the cases of extroversion, frequently cutting them off from their close neighbours when they paid allegiance to rival empires." Moreover, he states that, at the turn of the twentieth century, British cables "constituted two-thirds of the world network of underwater cables, and twenty-two of the twenty-five companies that managed international networks were British-affiliated." Mattelart, *Invention of Communication*, 170 and 167.

29. Mattelart, *Invention of Communication*, 161.

30. Mattelart, *Invention of Communication*, 202.

31. Forbes et al., *Northernmost Labrador Mapped from the Air*, 61.

32. Forbes et al., *Northernmost Labrador Mapped from the Air*, 62.

33. Forbes et al., *Northernmost Labrador Mapped from the Air*, 69.

34. Forbes et al., *Northernmost Labrador Mapped from the Air*, 167.

35. Berland, *North of Empire*, 74.

36. Berland also specifies the relationality between temporal and spatial "margins"

and the production of colonial space: "A 'margin' is a space which is drawn into the axes of imperial economy, administration, and information but which remains 'behind' (to put it in temporal terms) or 'outside' (spatially speaking) in terms of economic and political power. Communication technologies facilitate the simultaneous integration and extrusion of colonized territories. The margin is a spatial concept, but colonial space is the product, not the predecessor, of colonizing practices." Berland, *North of Empire*, 76.

37. Canguilhem, "Living and Its Milieu," 17.

38. The written Anglophone record has of course a severely biased representation of these living practices. A notable exception includes the narrative put down by John Igloliorte in *An Inuk Boy Becomes a Hunter*, which recounts his land-based youth in Nain, Labrador, during the 1940s and 1950s.

39. Canguilhem, "Living and Its Milieu," 17.

40. Forbes et al., *Northernmost Labrador Mapped from the Air*, 184.

41. Box 42, folder 1523, Re: Completion of the aerial survey of northern Labrador, 1936, Alexander Forbes Papers.

42. Forbes et al., *Northernmost Labrador Mapped from the Air*, 184–85.

43. Kember and Zylinska, *Life after New Media*, xv; phrase is italicized in the original.

44. O. M. Miller, "Ground Survey Report," box 100, folder 2187, Labrador Expedition, 1931, O. M. Miller's ground survey report, Alexander Forbes Papers.

45. Sterne, "Compression," 35.

46. Gilbert Simondon, quoted in Sterne, "Compression," 35.

47. Gilbert Simondon, quoted in Sterne, "Compression," 36.

48. Gilbert Simondon, quoted in Sterne, "Compression," 34.

49. For a contemporaneous history of extra-metropolitan aerial photography in Canada, see Bocking, "Disciplined Geography"; and McGrath and Sebert, *Mapping a Northern Land*.

50. Letter from Alexander Forbes to Isaiah Bowman, December 10, 1934, box 101, folder 2205, Labrador Expedition, Correspondence Jan. 1934–July 1935, Alexander Forbes Papers.

51. See Forbes, "Flight to Cape Chidley."

52. Forbes, *Quest for a Northern Air Route*.

53. Fenn, "Alexander Forbes, 1882–1965," 127–28. See Farish, "Frontier Engineering."

54. Parks and Starosielski, introduction, 14.

55. Nunatsiavut achieved self-government in 2005. For a remarkable articulation of land and cultural expression drawn out from Labrador, see Heather Igloliorte, *SakKijâjuk*. *SakKijâjuk* means "to be visible" in the dialect of Inukttut prevalent in Nunatsiavut.

56. Alexander Forbes, "Short-Cuts in Long-Distance Photogrammetry," box 46, folder 1573, Re: "Short-cuts in long-distance photogrammetry," 1944, Alexander Forbes Papers.

57. Forbes, "Photogrammetry Applied to Aerology," 181.

58. Hydrographic Office and Photographic Intelligence Center, "Introduction to Oblique Photogrammetry," foreword (n.p.), box 46, folder 1575, Re: "Introduction to oblique photogrammetry," 1945 (no. 1 of 5 folders), Alexander Forbes Papers.

59. Hydrographic Office and Photographic Intelligence Center, "Introduction to Oblique Photogrammetry," 1.

60. Letter from G. S. Bryan, hydrographer, to C. O., Photo Interpretation School, U.S. Naval Air Station, Washington, DC, Subj: Manual of Oblique Photogrammetry, September 18, 1944, box 46, folder 1576, Re: "Introduction to oblique photogramme-try," 1945 (no. 2 of 5 folders), Alexander Forbes Papers.

61. Letter from C. H. Cox, Photographic Interpretation Center, to the hydrographer, September 26, 1944, box 46, folder 1576, Re: "Introduction to oblique photogramme-try," 1945 (no. 2 of 5 folders), Alexander Forbes Papers.

62. Letter from Alexander Forbes to Robert Singleton, August, 1, 1946, box 112, folder 2375, World War II Notes Re: Bikini June–July 1946, Alexander Forbes Papers.

63. Letter from Alexander Forbes to editor of *Washington Post*, September 11, 1946, box 112, folder 2375, World War II Notes Re: Bikini June–July 1946, Alexander Forbes Papers.

64. Letter from Alexander Forbes to editor of *Washington Post*, September 11, 1946.

65. Letter from Alexander Forbes to editor of *Washington Post*, September 11, 1946.

66. Letter from Alexander Forbes to editor of *Washington Post*, September 11, 1946.

CHAPTER 4. THE PROMISE OF EXTRACTION

1. See King, *White Thunder*.

2. For an eyewitness account of the explosion by the ship's wireless operator, see King, "Viking's Last Cruise."

3. Newfoundland and Labrador Heritage, "Early Days of Film."

4. "A Talking Movie of the Mission," 88.

5. The DeVry Corporation manufactured early point-and-shoot film equipment for mass consumption from the 1910s to the 1950s.

6. King, *White Thunder*.

7. "Talking Movie of the Mission," 88.

8. "Talking Movie of the Mission," 88.

9. See Gibson-Graham, "Diverse Economies," and *Postcapitalist Politics*.

10. Peters, "History as a Communication Problem," 22.

11. See King, *Networked Reenactments*.

12. Tsing, *Mushroom at the End of the World*, 159–61.

13. Van Wyck, *Highway of the Atom*, 19.

14. Cameron, *Far Off Metal River*, 12.

15. Stewart, *Ordinary Affects*, 4.

16. See Bolter and Grusin, *Remediation*; and Acland, *Residual Media*.

17. Mulvey, *Death 24× a Second*, 9.

18. Mulvey, *Death 24× a Second*, 11.

19. Mulvey, *Death 24× a Second*, 10.

20. Mulvey, *Death 24× a Second*, 30.

21. Heise, "Unnatural Ecologies," 168.

22. Letter from Anthony Gardner (Grenfell Association of America) to Katie Spald-ing (Grenfell Association of Great Britain and Ireland), February 24, 1930, MG 63.64,

Lanterns Slides and Films 11 January 1928–4 April 1935, Grenfell Association of Great Britain and Ireland Papers.

23. Letter from Betty Sandbrook (Grenfell Association of Great Britain and Ireland) to IGA Directors (Grenfell Association of America), February 11, 1949, MG 63.1988, film of Sir Wilfred's life 1940, Grenfell Association of Great Britain and Ireland Papers.

24. "The Labrador Film Company, Inc." investor advertising sheet, MG 63.311, Grenfell Association of Great Britain and Ireland Papers.

25. The partnership dissolved due to the accidental death of Varick Frissell in 1931 during filming of *The Viking* off the coast of Newfoundland. It is also worth noting to what extent Flaherty saw his Grenfell Mission motion picture fitting in with his two previous films, *Nanook of the North* (1922) and *Moana of the South Sea* (1926). As the company's investor advertising sheet states, "He considers these stepping-stones to the great picture he wants to make in Labrador. One fact is certain, that because the story is not about eskimos or other primitives, but about white people working under hardships for a great cause in Labrador, the film will have 'human interest,' and therefore appeal to the public far more than 'Nanook' or 'Moana.'" See "The Labrador Film Company," MG 63.311, Grenfell Association of Great Britain and Ireland Papers.

26. "Labrador Film Company."

27. Cameron, *Far Off Metal River*, 25.

28. Cameron, *Far Off Metal River*, 25.

29. Stoler, *Along the Archival Grain*, 4.

30. See, by way of example, Alfred, *Wasáse*; Byrd, *Transit of Empire*; Cameron, *Far Off Metal River*; Ingersoll, *Waves of Knowing*; Smith, *Decolonizing Methodologies*.

31. Smith, *Decolonizing Methodologies*, 4.

32. Berlant, "Commons," 393.

33. Cowen, "Infrastructures of Empire and Resistance."

34. Cameron, *Far Off Metal River*, 20.

THE WAY IT WAS, ST. ANTHONY, 1997

1. Ernest Simms, personal interview, November 21, 2011, St. Anthony

2. St. Anthony Basin Resources Inc. (SABRI), "Background."

3. SABRI, "Background." In April 2013 SABRI and Clearwater Seafoods negotiated a ten-year extension to their contract for harvesting the allocation of northern shrimp.

4. SABRI, "St. Anthony Seafoods."

5. Curtis Richards, personal interview, November 21, 2011, St. Anthony.

6. Simms, personal interview.

7. Richards, personal interview.

8. Berlant, *Cruel Optimism*, 24.

9. Simms, personal interview.

MANUSCRIPTS

Canada

Captain George Mack Fonds, McCord Museum and Archives, Montreal, Quebec
Frederick Berchem Fonds, McCord Museum and Archives, Montreal, Quebec
Governors' Papers, Provincial Archives of Newfoundland and Labrador, St. John's,
Newfoundland and Labrador
Grenfell Association of Great Britain and Ireland Papers, Provincial Archives of New-
foundland and Labrador, St. John's, Newfoundland and Labrador
Hugh Peck Fonds, McCord Museum and Archives, Montreal, Quebec
International Grenfell Association Fonds, Provincial Archives of Newfoundland and
Labrador, St. John's, Newfoundland and Labrador
Labrador Institute Archives, Labrador Institute, Memorial University, Happy Valley-
Goose Bay, Newfoundland and Labrador
Labrador Medical Mission (Ottawa) Papers, Provincial Archives of Newfoundland and
Labrador, St. John's, Newfoundland and Labrador

United Kingdom

British Postal Museum and Archives, London
Dominions' Office Papers, National Archives, Kew, London

United States

Alexander Forbes Papers, Francis A. Countway Library of Medicine, Center for the His-
tory of Medicine, Harvard Medical Library, Cambridge, Massachusetts
Delano and Aldrich Architectural Records and Papers, Avery Architectural and Fine
Arts Library, Columbia University, New York
New England Grenfell Association Records, Sterling Library, Yale University, New Ha-
ven, Connecticut
Wilfred Thomason Grenfell Papers, Sterling Library, Yale University, New Haven,
Connecticut
William Adams Delano Papers, Sterling Library, Yale University, New Haven,
Connecticut
William Adams Delano Reminiscences, Oral History Project, Columbia University,
New York

PUBLISHED SOURCES

Acland, Charles, ed. *Residual Media*. Minneapolis: University of Minnesota Press, 2006.

Adams, Annmarie. *Architecture in the Family Way: Doctors, Houses, Women*. Montreal: McGill-Queen's University Press, 1996.

Adams, Annmarie. "Modernism and Medicine: The Hospitals of Stevens and Lee, 1916–1932." *Journal of the Society of Architectural Historians* 58, no. 1 (1999): 42–61.

Addams, Jane. *Twenty Years at Hull House*. New York: Macmillan, 1911.

Alfred, Taiaiake. *Wasáse: Indigenous Pathways of Action and Freedom*. Peterborough, ON: Broadview Press, 2005.

Anderson, Warwick. *Colonial Pathologies: American Tropical Medicine, Race, and Hygiene in the Philippines*. Durham, NC: Duke University Press, 2006.

Appel, Hannah, Nikhil Anand, and Akhil Gupta. "Temporality, Politics, and the Promise of Infrastructure." In *The Promise of Infrastructure*, edited by Nikhil Anand, Akhil Gupta, and Hannah Appel, 1–38. Durham, NC: Duke University Press, 2018.

"Area of Newfoundland Trebled by Privy Council Ruling." *Among the Deep Sea Fishers* 25, no. 1 (April 1927): 10.

Barney, Darin. "To Hear the Whistle Blow: Technology and Politics on the Battle River Branch Line." *TOPIA: Canadian Journal of Cultural Studies*, no. 25 (2011): 5–28.

Barry, Andrew. "Discussion: Infrastructural Times." *Cultural Anthropology*, September 24, 2015. https://culanth.org/fieldsights/discussion-infrastructural-times.

Bavington, Dean. *Managed Annihilation: An Unnatural History of the Newfoundland Cod Collapse*. Vancouver: University of British Columbia Press, 2010.

Berland, Jody. *North of Empire: Essays on the Cultural Technologies of Space*. Durham, NC: Duke University Press, 2009.

Berlant, Lauren. "The Commons: Infrastructures for Troubling Times." *Environment and Planning D: Society and Space* 34, no. 3 (2016): 393–419.

Berlant, Lauren. *Cruel Optimism*. Durham, NC: Duke University Press, 2011.

Blackburn, A. C. "How Stores and Supplies Are Handled at St. Anthony." *Among the Deep Sea Fishers* 25, no. 2 (July 1927): 65.

Bocking, Stephen. "A Disciplined Geography: Aviation, Science, and the Cold War in Northern Canada, 1945–1960." *Technology and Culture* 50, no. 2 (2008): 265–90.

Bolter, Jay D., and Richard Grusin. *Remediation: Understanding New Media*. Cambridge, MA: MIT Press, 1999.

Bonnett, John. *Empire and Emergence: Innis, Complexity, and the Trajectory of History*. Montreal: McGill-Queen's University Press, 2013.

Bowker, Geoffrey. "Information Mythology and Infrastructure." In *Information Acumen: The Understanding and Use of Knowledge in Modern Business*, edited by Lisa Bud-Frierman, 231–41. London: Routledge, 1994.

Bowker, Geoffrey. *Science on the Run: Information Management and Industrial Geophysics at Schlumberger, 1920–1940*. Cambridge, MA: MIT Press 1994.

Bozak, Nadia. *The Cinematic Footprint: Lights, Cameras, Natural Resources*. New Brunswick, NJ: Rutgers University Press, 2011.

Briggs, Asa. *Victorian Things*. Rev. ed. Stroud, UK: Sutton, 2003.

Brown, Stephen, ed. *Struggling for Effectiveness: CIDA and Canadian Foreign Aid*. Montreal: McGill-Queen's University Press, 2012.

Burgess, C. Perry. "The Campaign." *Among the Deep Sea Fishers* 19, no. 1 (April 1921): 6–7.

Buxton, William, ed. *Harold Innis and the North: Appraisals and Contestations*. Montreal: McGill-Queen's University Press, 2013.

Buxton, William, and Charles Acland, eds. *Harold Innis in the New Century: Reflections and Refractions*. Montreal: McGill-Queen's University Press, 1999.

Byrd, Jodi. *The Transit of Empire: Indigenous Critiques of Colonialism*. Minneapolis: University of Minnesota Press, 2011.

Cadigan, Sean. "Battle Harbour in Transition: Merchants, Fishermen and the State in the Struggle for Relief in a Labrador Community during the 1930s." *Labour /Le Travail* 26 (Fall 1990): 125–50.

Cadigan, Sean. *Hope and Deception in Conception Bay: Merchant-Settler Relations in Newfoundland, 1785–1855*. Toronto: University of Toronto Press, 1995.

Cahill, Edward. "Brock Process of Aerial Mapping." *Journal of the Optical Society of America* 22, no. 3 (1932): 111–36.

Cameron, Emilie. *Far Off Metal River: Inuit Lands, Settler Stories, and the Making of the Contemporary Arctic*. Vancouver: University of British Columbia Press, 2015.

Cameron, Emilie, Sarah de Leeuw, and Margo Greenwood. "Indigeneity." In *International Encyclopedia of Human Geography*, 5th ed., edited by Rob Kitchin and Nigel Thrift, 352–57. London: Elsevier, 2009.

Canguilhem, Georges. "The Living and Its Milieu." Translated by John Savage. *Grey Room* 3 (Spring 2001): 6–31.

Carse, Ashley. *Beyond the Big Ditch: Politics, Ecology, and Infrastructure at the Panama Canal*. Cambridge, MA: MIT Press, 2014.

Carson, Mina. *Settlement Folk: Social Thought and the American Settlement Movement, 1885–1930*. Chicago: University of Chicago Press, 1990.

Chow, Rey. *The Age of the World Target: Self-Referentiality in War, Theory, and Comparative Work*. Durham, NC: Duke University Press, 2006.

Connolly, Cynthia. *Saving Sickly Children: The Tuberculosis Preventorium in American Life, 1909–1970*. New Brunswick, NJ: Rutgers University Press, 2008.

Coombs-Thorne, Heidi. "Conflict and Resistance to Paternalism: Nursing with the Grenfell Mission Stations in Newfoundland and Labrador, 1939–81." In *Caregiving on the Periphery: Historical Perspectives on Nursing and Midwifery in Canada*, edited by Myra Rutherdale, 210–42. Montreal: McGill-Queen's University Press, 2011.

"Co-operative Stores among Atlantic Fishermen." *Among the Deep Sea Fishers* 1, no. 2 (July 1903): 9.

Cosby, Arthur. "Impressions of the Mission Stations." *Among the Deep Sea Fishers* 21, no. 3 (October 1923): 75–81.

Cosby, Arthur. "A New Hospital for St. Anthony." *Among the Deep Sea Fishers* 23, no. 1 (April 1925): 2–3.

Cowen, Deborah. "Infrastructures of Empire and Resistance." Verso Blog, January 25, 2017. http://www.versobooks.com/blogs/3067-infrastructures-of-empire-and-resistance.

Crellin, John. *Home Medicine: The Newfoundland Experience*. Montreal: McGill-Queen's University Press, 1994.

Curtis, Charles S. "Activities at St. Anthony in 1934." *Among the Deep Sea Fishers* 33, no. 1 (April 1935): 3–14.

Curtis, Charles S. "St. Anthony." *Among the Deep Sea Fishers* 22, no. 4 (January 1925): 162–63.

Curtis, Charles S. "St. Anthony Hospital." *Among the Deep Sea Fishers* 33, no. 1 (April 1925): 26.

Curtis, Charles S. "The Year's Work at St. Anthony Hospital." *Among the Deep Sea Fishers* 23, no. 4 (January 1926): 147–49.

Dafnos, Tia, and Shiri Pasternak. "How Does a Settler State Secure the Circuitry of Capital?" *Environment and Planning D: Society and Space* 36, no. 4 (2018): 739–57.

Davis, Howard. *The Culture of Building*. New York: Oxford University Press, 2006.

Demarest, Emma. "History of St. Anthony Hospital." *Among the Deep Sea Fishers* 22, no. 3 (October 1925): 103–7.

Dictionary of Canadian Biography Online. "Grenfell, Sir Wilfred Thomason." Accessed February 18, 2014. http://www.biographi.ca/en/bio/grenfell_wilfred_thomason_16E.html.

Drache, Daniel. "Introduction: Celebrating Innis: The Man, the Legacy, and Our Future." In *Staples, Markets, and Cultural Change: Selected Essays of Harold Innis*, edited by Daniel Drache. Montreal: McGill-Queen's University Press, 1995.

Dreicer, Gregory. "Building Bridges and Boundaries: The Lattice and the Tube, 1820–1860." *Technology and Culture* 51, no. 1 (2010): 126–63.

Duncan, Norman. *Dr. Grenfell's Parish: The Deep Sea Fishermen*. Toronto: F. H. Revell, 1905.

Edwards, Paul. "Infrastructure and Modernity: Force, Time, and Social Organization in the History of Sociotechnical Systems." In *Modernity and Technology*, edited by Thomas J. Misa, Philip Brey, and Andrew Feenberg, 185–225. Cambridge, MA: MIT Press, 2003.

Evenden, Matthew. "The Northern Vision of Harold Innis." *Journal of Canadian Studies* 34, no. 3 (1999): 162–86.

Farish, Matthew. "Frontier Engineering: From the Globe to the Body in the Cold War Arctic." *Canadian Geographer* 50, no. 2 (2006): 177–96.

Fenn, Wallace O. "Alexander Forbes, 1882–1965: A Biographical Memoir." Washington, DC: National Academy of Sciences, 1969.

Fishermen's Mission. "Our Mission." Accessed February 18, 2014. http://www.fishermensmission.org.uk/index.php?page=what-we-do.

Forbes, Alexander. "A Flight to Cape Chidley, 1935." *Geographical Review* 26, no. 1 (1936): 48–58.

Forbes, Alexander. "The Mechanism of Reaction." In *The Foundations of Experimental Psychology*, edited by Carl Murchison, 128–61. Worcester, MA: Clark University Press, 1929.

Forbes, Alexander. "Photogrammetry Applied to Aerology." *Photogrammetric Engineering* 11 (July, August, September 1945): 181–92.

Forbes, Alexander. *Quest for a Northern Air Route*. Cambridge, MA: Harvard University Press, 1953.

Forbes, Alexander. "Sir Wilfred's Vision." *Among the Deep Sea Fishers* 39, no. 4 (January 1942): 114–15.

Forbes, Alexander, with O. M. Miller, N. E. Odell, and Ernst C. Abbe. *Northernmost Labrador Mapped from the Air*. New York: American Geographical Society, 1938.

"Frontispiece." *Among the Deep Sea Fishers* 25, no. 2 (July 1927): 49.

Gabrys, Jennifer. *Digital Rubbish: A Natural History of Electronics*. Ann Arbor: University of Michigan Press, 2013.

Gabrys, Jennifer. *Program Earth: Environmental Sensing Technology and the Making of a Computational Planet*. Minneapolis: University of Minnesota Press, 2016.

Galloway, Alexander R., Eugene Thacker, and McKenzie Wark. *Excommunication: Three Inquiries in Media and Mediation*. Chicago: University of Chicago Press, 2014.

Gibson-Graham, J. K. "Diverse Economies: Performative Practices for 'Other Worlds.'" *Progress in Human Geography* 32 (2008): 613–32.

Gibson-Graham, J. K. *A Postcapitalist Politics*. Minnesota: University of Minnesota Press, 2006.

Gitelman, Lisa. *Always Already New: Media, History and the Data of Culture*. Cambridge, MA: MIT Press, 2006.

Gitelman, Lisa. *Paper Knowledge: Toward a Media History of Documents*. Durham, NC: Duke University Press, 2014.

Gómez-Barris, Macarena. *The Extractive Zone: Social Ecologies and Decolonial Perspectives*. Durham, NC: Duke University Press, 2017.

Government of Newfoundland and Labrador. "A New Dawn for the Labrador Innu," News release, November 18, 2011. http://www.releases.gov.nl.ca/releases/2011/exec/1118n11.htm.

Graeber, David. *Debt: The First 5,000 Years*. Brooklyn: Melville House, 2011.

Greene, Theodore Ainsworth. "The Opening of St. Anthony Hospital." *Among the Deep Sea Fishers* 25, no. 3 (October 1927): 99–106.

Greene, Theodore Ainsworth. "William Adams Delano." *Among the Deep Sea Fishers* 25, no. 2 (July 1927): 64.

Grenfell Interpretation Centre. Grenfell Handicrafts Online Store. Accessed May 27, 2020. http://www.grenfell-properties.com/handicrafts.

Grenfell, Wilfred. *Adrift on an Icepan*. Boston: Houghton Mifflin, 1909.

Grenfell, Wilfred. "Crossing the Sagebrush Prairies." *Among the Deep Sea Fishers* 19, no. 1 (April 1921): 5–6.

Grenfell, Wilfred. "Doctor Grenfell's Log." *Among the Deep Sea Fishers* 11, no. 1 (April 1913): 3–11.

Grenfell, Wilfred. *Down to the Sea*. New York: Fleming H. Revell Co., 1910.

Grenfell, Wilfred. "Dr. Grenfell Writes of the Holy Land." *Among the Deep Sea Fishers* 23, no. 1 (April 1925): 4–7.

Grenfell, Wilfred. *Forty Years for Labrador*. Boston: Houghton Mifflin, 1932.

Grenfell, Wilfred. "'Grenfell Cloth' for Arctic and Tropic Wear." *Among the Deep Sea Fishers* 26, no. 1 (April 1928): 34–35.

Grenfell, Wilfred. *A Labrador Doctor: The Autobiography of Wilfred Thomason Grenfell, M.D. (Oxon.), C.M.G.* Boston: Houghton Mifflin, 1919; London: Hodder and Stoughton, 1920.

Grenfell, Wilfred. *Labrador Looks at the Orient: Notes of Travel in the Near and Far East.* Boston: Houghton Mifflin, 1928.

Grenfell, Wilfred. *Labrador's Fight for Economic Freedom.* Self and Society, no. 19. London: Ernest Benn Ltd., 1929.

Grenfell, Wilfred. "The Log of the S.S. Strathcona—1926." *Among the Deep Sea Fishers* 24, no. 3 (October 1926): 112–17.

Grenfell, Wilfred. "The Lure of the Labrador." *National Review*, no. 97 (1931): 623–28.

Grenfell, Wilfred. "Movement to Perpetuate the Work of Dr. Grenfell." *Among the Deep Sea Fishers* 18, no. 3 (October 1920): 99–101.

Grenfell, Wilfred. "The New Hospital for St. Anthony." *Among the Deep Sea Fishers* 23, nos. 2 and 3 (July and October 1925): 51–52, 99–103.

Grenfell, Wilfred. "The Problems of Labrador." *Canadian Geographical Journal* 7, no. 5 (November 1933): 200–212.

Grenfell, Wilfred. *Religion in Everyday Life.* Chicago: American Library Association, 1926.

Grenfell, Wilfred. *The Romance of Labrador.* New York: Macmillan, 1934.

Grenfell, Wilfred. "Sir Wilfred's Log: Summer of 1934." *Among the Deep Sea Fishers* 32, no. 3 (October 1934): 93–104.

Grenfell, Wilfred. "To What Church Do You Belong?" *Among the Deep Sea Fishers* 27, no. 4, (January 1920): 116–17.

Grenfell, Wilfred. *Vikings of Today; or, Life and Medical Work among the Fishermen of Labrador.* New York: Fleming H. Revell Co., 1895.

Grenfell, Wilfred. *What Can Jesus Christ Do with Me?* Boston: Pilgrim Press, 1912.

Grenfell, Wilfred. *What the Church Means to Me: A Frank Confession and a Friendly Estimate by an Insider by Wilfred T. Grenfell, M.D. (Oxon.) Superintendent Labrador Medical Mission.* Boston: Pilgrim Press, 1911.

Grenfell, Wilfred. "Who Will Provide for These Special Needs?" *Among the Deep Sea Fishers* 13, no. 2 (July 1915): 54–55.

Grenfell Interpretation Centre. Grenfell Handicrafts Online Store. Accessed May 27, 2020. http://www.grenfell-properties.com/handicrafts.

Grusin, Richard. "Radical Mediation." *Critical Inquiry* 42, no. 1 (2015): 124–48.

Guillory, John. "Genesis of the Media Concept." *Critical Inquiry* 36, no. 2 (2010): 321–62.

Handy, Robert T. *A Christian America: Protestant Hopes and Historical Realities.* New York: Oxford University Press, 1971.

Handy, Robert T. *The Social Gospel in America, 1870–1920.* New York: Oxford University Press, 1966.

Hause, Frank E. "The Design and Installation of an Electric Lighting Plant for Dr. W. T. Grenfell." In *Science and Technology Annual*, 55–64. Brooklyn: Pratt Institute, 1909.

Hayles, N. Katherine. "Searching for Common Ground." In *Reinventing Nature? Re-*

sponses to Postmodern Deconstruction, edited by Michael E. Soulé and Gary Leas, 47–63. Washington, DC: Island Press, 1995.

Hayles, N. Katherine. "Simulated Nature and Natural Simulations: Rethinking the Relation between the Beholder and the World." In *Uncommon Ground: Rethinking the Human Place in Nature*, edited by William Cronon, 409–25. New York: Norton, 1995.

Heise, Ursula. "Unnatural Ecologies: The Metaphor of the Environment in Media Theory." *Configurations* 10, no. 1 (2002): 149–68.

Hiller, James. "Grenfell and His Successors." *Newfoundland and Labrador Studies* 10, no. 1 (1994): 124–31.

Hiller, James. "The Newfoundland Credit System: An Interpretation." In *Merchant Credit and Labour Strategies*, edited by Rosemary Ommer, 86–101. Fredericton, NB: Acadiensis, 1990.

Hiller, James, and Peter Neary, eds. *Newfoundland in the Nineteenth and Twentieth Centuries*. Toronto: University of Toronto Press, 1980.

Hilton, George Woodman. *The Truck System, including a History of the British Truck Acts, 1465–1960*. Cambridge: W. Heffer, 1960.

Hobday, R. A. "Sunlight Therapy and Solar Architecture." *Medical History* 41, no. 4 (1997): 455–72.

Hoffman, Beatrix Rebecca. *Health Care for Some: Rights and Rationing in the United States since 1930*. Chicago: University of Chicago Press, 2012.

Holloway, Robert E. "Preface." In *Through Newfoundland with the Camera*. St. John's, NL: Dicks and Co., 1905.

Hopkins, Charles Howard. *The Rise of the Social Gospel in American Protestantism*. New Haven, CT: Yale University Press, 1940.

Hornsby, John A. "The Trend of Modern Hospital Service." *Among the Deep Sea Fishers* 13, no. 3 (October 1915): 97–98.

Hornsby, John A., and Richard E. Schmidt. *The Modern Hospital*. Philadelphia: W. B. Saunders Co., 1913.

Howell, Joel. *Technology in the Hospital: Transforming Patient Care in the Early Twentieth Century*. Baltimore, MD: Johns Hopkins University Press, 1995.

Humber, Elizabeth A. "The Shape of Authenticity in Elliot's Northern Nurse." *Journal of Newfoundland Studies* 23, no. 1 (2008): 45–59.

Igloliorte, Heather, ed. *SakKijâjuk: Art and Craft from Nunavut*. Fredericton, NB: Goose Lane Editions and the Rooms Corp., 2017.

Igloliorte, John. *An Inuk Boy Becomes a Hunter*. Halifax, NS: Nimbus, 1994.

Ingersoll, Karin Amimoto. *Waves of Knowing: A Seascape Epistemology*. Durham, NC: Duke University Press, 2017.

Innis, Harold. *The Bias of Communication*. Toronto: University of Toronto Press, 1951.

Innis, Harold. *The Cod Fisheries: The History of an International Economy*. New Haven, CT: Yale University Press, 1940.

Innis, Harold. *Empire and Communications*. Oxford: Clarendon, 1950.

Innis, Harold. *The Fur Trade in Canada: An Introduction to Canadian Economic History*. Toronto: University of Toronto Press, 1930.

Innis, Harold. *A History of the Canadian Pacific Railway*. Toronto: University of Toronto Press, 1923.

Jackson, Gregory S. "Cultivating Spiritual Sight: Jacob Riis's Virtual-Tour Narrative and the Visual Modernization of Protestant Homiletics." *Representations* 83, no. 1 (2003): 126–66.

Jackson, Gregory S. *The Word and Its Witness: The Spiritualization of American Realism*. Chicago: University of Chicago Press, 2009.

Jackson, Steven J. "Rethinking Repair." In *Media Technologies: Essays on Communication, Materiality and Society*, edited by Tarleton Gillespie, Pablo J. Boczkowski, and Kirsten A. Foot, 221–40. Cambridge, MA: MIT Press, 2014.

Jones-Imhotep, Edward. *The Unreliable Nation: Hostile Nature and Technological Failure in the Cold War*. Cambridge, MA: MIT Press, 2017.

Jue, Melody. *Wild Blue Media: Thinking through Seawater*. Durham, NC: Duke University Press, 2020.

Kember, Sarah, and Joanna Zylinska. *Life after New Media: Mediation as a Vital Process*. Cambridge, MA: MIT Press, 2012.

Kerr, J. Lennox. *Wilfred Grenfell: His Life and Work*. Toronto: Ryerson, 1959.

King, Clayton L. "The Viking's Last Cruise." *Evening Telegram* (St. John's), 1935. Memorial University Centre for Newfoundland Studies, Digital Archives Initiative. Accessed March 10, 2014, http://collections.mun.ca/cdm4/document.php ?CISOROOT=/cns&CISOPTR=102398&REC=4.

King, Katie. "A Naturalcultural Collection of Affections: Transdisciplinary Stories of Transmedia Ecologies Learning." *Scholar and Feminist Online* 10, no. 3 (Summer 2012).

King, Katie. *Networked Reenactments: Stories Transdisciplinary Knowledges Tell*. Durham, NC: Duke University Press, 2011.

King, Victoria, dir. *White Thunder: The Story of Varick Frissell and the* Viking *Disaster*. Harrington Park, NJ: Milestone Entertainment, 2004.

Kittler, Friedrich. *Gramophone-Film-Typewriter*. Translated by Geoffrey Winthrop-Young and Michael Wutz. Stanford, CA: Stanford University Press, 1999.

Kurgan, Laura. *Close Up at a Distance: Mapping, Technology, and Politics*. New York: Zone Books, 2013.

Labrador Resources Advisory Council. "Labrador Conference, Tape 1." Memorial University of Newfoundland, Digital Archives Initiative. Accessed February 18, 2014. http://collections.mun.ca/cdm4/item_viewer.php?CISOROOT=/&CISOPTR=1742 &CISOBOX=1&REC=1.

Laverty, Paula. *Silk Stocking Mats: Hooked Mats of the Grenfell Mission*. Montreal: McGill-Queen's University Press, 2005.

Lazzarato, Maurizio. *The Making of the Indebted Man*. Amsterdam: Semiotext(e), 2011.

Macdonald, Ian D. *"To Each His Own": William Coaker and the Fishermen's Protective Union in Newfoundland Politics, 1908–1925*. St. John's, NL: Institute for Social and Economic Research, 1987.

Mackinnon, Paul. *Vernacular Architecture in the Codroy Valley*. Hull, QC: Canadian Museum of Civilization, 2002.

"Many Mansions." *Time*, November 12, 1928, 34.

Mattelart, Armand. *The Invention of Communication*. Translated by Susan Emanuel. Minneapolis: University of Minnesota Press, 1996.

Mattern, Shannon. *Deep Mapping the Media City*. Minneapolis: University of Minnesota Press, 2015.

Mattern, Shannon. "Infrastructural Tourism." *Places Journal* (July 2013). http://placesjournal.org/article/infrastructural-tourism

Maxwell, Richard, and Toby Miller. *Greening the Media*. New York: Oxford University Press, 2012.

Mayou, Edith. "Harrington Letter." *Among the Deep Sea Fishers* 6, no. 4 (January 1909): 33–36.

McGeer, Michael. *A Fierce Discontent: The Rise and Fall of the Progressive Movement in America, 1870–1920*. New York: Free Press, 2003.

McGrath, Gerald, and Louis Sebert, eds. *Mapping a Northern Land: The Survey of Canada, 1947–1994*. Montreal: McGill-Queen's University Press, 1999.

McKay, Ian. *The Quest of the Folk: Antimodernism and Cultural Selection in Twentieth-Century Nova Scotia*. Montreal: McGill-Queen's University Press, 1994.

McKegney, Sam W. "Second-Hand Shaman: Imag(in)ing Indigeneity from Le Jeune to Pratt, Moore and Beresford." *TOPIA: Canadian Journal of Cultural Studies* 12 (Fall 2004): 25–40.

McNaught, Kenneth. *A Prophet in Politics: A Biography of J. S. Woodsworth*. Toronto: University of Toronto Press, 1959.

Mellin, Robert. *Tilting: House Launching, Slide Hauling, Potato Trenching, and Other Tales from a Newfoundland Fishing Village*. New York: Princeton Architectural Press, 2003.

"The Men behind the Scenes." *Among the Deep Sea Fishers* 25, no. 2 (July 1927): 66.

Merrick, Elliot. *Northern Nurse*. New York: Scribner's, 1942.

Miller, O. M. "Planetabling from the Air: An Approximate Method of Plotting from Oblique Aerial Photographs." *Geographical Review* 21, no. 2 (1931): 201–12.

Mitchell, Helen. "Nutrition Survey in Labrador and Northern Newfoundland." *Journal of the American Dietetic Association* 6 (1930): 33.

Mitchell, W. J. T. *What Do Pictures Want? The Lives and Loves of Images*. Chicago: University of Chicago Press, 2004.

Mulvey, Laura. *Death 24× a Second: Stillness and the Moving Image*. London: Reaktion, 2006.

Naylor, C. David, ed. *Canadian Health Care and the State: A Century of Evolution*. Montreal: McGill-Queen's University Press, 1992.

Newell, I. "Credit and Common Sense." *Among the Deep Sea Fishers* 39, no. 3 (October 1941): 82–84.

Newfoundland and Labrador Heritage. "The Beothuk." Accessed June 6, 2020. https://www.heritage.nf.ca/articles/aboriginal/beothuk.php.

Newfoundland and Labrador Heritage. "Early Days of Film." Accessed June 13, 2020. http://www.heritage.nf.ca/arts/early_days.html#frissell.

Nixon, Rob. *Slow Violence and the Environmentalism of the Poor*. Cambridge, MA: Harvard University Press, 2011.

O'Brien, Patricia, ed. *The Grenfell Obsession: An Anthology*. St. John's, NL: Creative Publishers, 1992.

Ommer, Rosemary. "Merchant Credit and the Informal Economy, 1919–1929." *Canadian Historical Papers* (1989): 167–89.

Ommer, Rosemary. "The Truck System in Gaspé, 1822–1877." In *Merchant Credit and Labour Strategies in Historical Perspective*, edited by Rosemary Ommer, 42–73. Fredericton, NB: Acadiensis, 1990.

Ommer, Rosemary, ed. *Merchant Credit and Labour Strategies in Historical Perspective*. Fredericton, NB: Acadiensis, 1990.

"Our Job: Information." *Atlantic Guardian* 1, no. 1 (January 1945): 12.

Paddon, Harry. "Indian Harbour Hospital." *Them Days: Stories of Early Labrador* 4, no. 1 (1978): 10.

Paddon, Harry. "Items from Indian Harbour." *Among the Deep Sea Fishers* 11, no. 2 (July 1913): 18.

Parks, Lisa. *Culture in Orbit: Satellites and the Televisual*. Durham: Duke University Press, 2005.

Parks, Lisa. "Drones, Vertical Mediation, and the Targeted Class." *Feminist Studies* 42, no. 1 (2016): 227–35.

Parks, Lisa. *Rethinking Media Coverage: Vertical Mediation and the War on Terror*. New York: Routledge, 2018.

Parks, Lisa, and James Schwoch, eds. *Down to Earth: Satellite Technologies, Industries, and Cultures*. New Brunswick, NJ: Rutgers University Press, 2012.

Parks, Lisa, and Nicole Starosielski. Introduction to *Signal Traffic: Critical Studies of Media Infrastructures*, edited by Lisa Parks and Nicole Starosielski, 1–28. Urbana: University of Illinois Press, 2015.

Parrikka, Jussi, ed. *Medianatures: The Materiality of Information Technology and Electronic Waste*. Ann Arbor, MI: Open Humanities Press, 2011.

Parsons, Charles. "Notre Dame Bay Memorial Hospital." *Among the Deep Sea Fishers* 21, no. 4 (January 1924): 115–21.

Patey, Francis. *A Battle Lost*. St. Anthony: Patey's Publications, 1991.

Patey, Francis. *The Grenfell Dock*. St. Anthony: Bebb Publishing, 1993.

Patey, Francis. *The Jolly Poker*. St. John's, NL: H. Cuff, 1992.

Peters, John Durham. "History as a Communication Problem." In *Explorations in Communication and History*, edited by Barbie Zelizer, 19–34. London: Routledge, 2008.

Peters, John Durham. *The Marvelous Clouds: Toward a Philosophy of Elemental Media*. Chicago: University of Chicago Press, 2015.

Peters, John Durham. "Strange Sympathies: Horizons of Media Theory in America and Germany." In *American Studies as Media Studies*, edited by Frank Kelleter and Daniel Stein, 1–26. Heidelberg: Universitätsverlag Winter, 2008.

Pilgrim, Earl B. *The Captain and the Girl*. St. John's, NL: Flanker, 2001.

Pilgrim, Earl B. *The Day Grenfell Cried*. St. John's, NL: DRC Publishing, 2007.

Piper, Liza. *The Industrial Transformation of Subarctic Canada*. Vancouver: University of British Columbia Press, 2010.

"Pleasant Commendation for the Hospital." *Among the Deep Sea Fishers* 23, no. 4 (January 1926): 192.

Pocius, Gerald. *A Place to Belong: Community Order and Everyday Space in Calvert, Newfoundland*. Montreal: McGill-Queen's University Press, 2001.

Proceedings of the Conference on Labrador Affairs held Feb. 13th–16th, 1956. St. John's, NL: Guardian Ltd., 1956.

Prowse, D. W. *A History of Newfoundland from the English, Colonial, and Foreign Records*. London: Macmillan, 1895.

Ray, Arthur. *The Canadian Fur Trade in the Industrial Age*. Toronto: University of Toronto Press, 1990.

Redfern, Percy. *The Story of the C.W.S.: The Jubilee History of the Co-operative Wholesale Society Limited, 1863–1913*. Manchester, UK: Co-operative Wholesale Society Ltd., 1913.

Roberts, John. "Darwinism, American Protestant Thinkers, and the Puzzle of Motivation." In *Disseminating Darwinism: The Role of Place, Race, Religion, and Gender*, edited by Ronald L. Numbers and John Stenhouse, 142–72. Cambridge: Cambridge University Press, 1999.

Roberts, Peter J. "The Process of Change." *Among the Deep Sea Fishers* 78, no. 2 (July 1981): 1–4.

Rollman, Hans. *Labrador through Moravian Eyes: 250 Years of Art, Photographs and Records*. St. John's, NL: Special Celebrations Corp. of Newfoundland and Labrador, 2002.

Rompkey, Ronald. *Grenfell of Labrador: A Biography*. Toronto: University of Toronto Press, 1991.

Rompkey, Ronald, ed. *Jessie Luther at the Grenfell Mission*. Montreal: McGill-Queen's University Press, 2001.

Rompkey, Ronald, ed. and trans. *The Labrador Memoir of Dr. Harry Paddon, 1912–1938*. Montreal: McGill-Queen's University Press, 2003.

Rompkey, Ronald, ed. *Labrador Odyssey: The Journal and Photographs of Eliot Curwen on the Second Voyage of Wilfred Grenfell, 1893*. Montreal: McGill-Queen's University Press, 1996.

Rudofsky, Bernard. *Architecture without Architects: An Introduction to Nonpedigreed Architecture*. New York: Museum of Modern Art, 1964.

Ruiz, Rafico. "Behind the Scenes at the Grenfell Mission: Edgar 'Ted' McNeil and Counter-Biography as Material Agency." In *The Grenfell Medical Mission and American Support in Newfoundland and Labrador, 1890s–1940s*, edited by Jennifer Connor and Katherine Side, 220–44. Montreal: McGill-Queen's University Press, 2018.

Ruiz, Rafico. "Grenfell Cloth: Weatherproof Textiles, Agency, and Historiographical Artefactuality." In *New Materials: Their Social and Cultural Meanings*, edited by Amy E. Slaton, 238–70. Amherst, MA: Lever, 2020.

Ryan, Shannon. *The Ice Hunters: A History of Newfoundland Sealing to 1914*. St. John's, NL: Breakwater Books, 1994.

Saint-Amour, Paul K. "Applied Modernism: Military and Civilian Uses of the Aerial Photomosaic." *Theory, Culture and Society* 28, nos. 7–8 (2011): 241–69.

Sharma, Jayeeta. *Empire's Garden: Assam and the Making of India*. Durham, NC: Duke University Press, 2011.

Sider, Gerald. *Culture and Class in Anthropology and History: A Newfoundland Illustration*. Cambridge: Cambridge University Press, 1986.

"Sixth Annual Report of the International Grenfell Association." *Among the Deep Sea Fishers* 28, no. 2 (July 1920): 76–96.

Smallwood, Joseph R., and Robert D. W. Pitt, eds. "Margaret Digby." In *Encyclopedia of Newfoundland and Labrador*, 1:623. St. John's, NL: Newfoundland Book Publishers, 1981.

Smith, Linda Tuhiwai. *Decolonizing Methodologies: Research and Indigenous Peoples*. London: Zed, 1999.

Sobchack, Vivian. "Afterword: Media Archaeology and Re-presencing the Past." In *Media Archaeology: Approaches, Applications, and Implications*, edited by Erkki Huhtamo and Jussi Parrikka, 323–33. Berkeley: University of California Press, 2011.

Spencer, Herbert. *The Principles of Biology*. New York: D. Appleton and Co., 1898.

Spice, Anne. "Fighting Invasive Infrastructures: Indigenous Relations against Pipelines." *Environment and Society* 9, no. 1 (2018): 40–56.

St. Anthony Basin Resources, Inc. (SABRI). "Background." SABRI website. Accessed June 14, 2020. http://www.sabrinl.com/Background.

St. Anthony Basin Resources, Inc. (SABRI). "St. Anthony Seafoods." SABRI website. Accessed June 14, 2020. http://sabrinl.com/st-anthony-seafoods.

Star, Susan Leigh. "The Ethnography of Infrastructure." *American Behavioral Scientist* 43, no. 3 (1999): 377–91.

Star, Susan Leigh, and Karen Ruhleder. "Steps toward an Ecology of Infrastructure: Design and Access for Large-Scale Information Spaces." *Information Systems Research* 7, no. 1 (1996): 111–34.

Starosielski, Nicole. *The Undersea Network*. Durham, NC: Duke University Press, 2015.

Sterne, Jonathan. "Compression: A Loose History." In *Signal Traffic: Critical Studies of Media Infrastructures*, edited by Lisa Parks and Nicole Starosielski, 31–52. Champaign: University of Illinois Press, 2015.

Sterne, Jonathan. *MP3: The Meaning of a Format*. Durham, NC: Duke University Press, 2012.

Sterne, Jonathan. "Transportation and Communication: Together as You've Always Wanted Them." In *Thinking with James Carey: Essays on Communications, Transportation, History*, edited by Jeremy Packer and Craig Robertson, 117–36. New York: Peter Lang, 2006.

Stewart, Kathleen. *Ordinary Affects*. Durham, NC: Duke University Press, 2007.

Stoler, Ann Laura. *Along the Archival Grain: Epistemic Anxieties and Colonial Common Sense*. Princeton, NJ: Princeton University Press, 2010.

Stoler, Ann Laura. *Duress: Imperial Durabilities in Our Times*. Durham, NC: Duke University Press, 2016.

"A Talking Movie of the Mission." *Among the Deep Sea Fishers* 26, no. 2 (July 1928): 88.

Tallant, Edith. *The Girl Who Was Marge*. Philadelphia: J. B. Lippincott, 1939.

Thomas, Gordon. "Farewell and Godspeed." *Among the Deep Sea Fishers* 78, no. 2 (July 1981): 13–15.

Thomas, Gordon. *From Sled to Satellite: My Years with the Grenfell Mission.* Toronto: Irwin, 1987.

Threlkeld-Edwards, Herbert. "The New Medical Era in St. Anthony." *Among the Deep Sea Fishers* 25, no. 2 (July 1927): 51–57.

Toland, Harry. *A Sort of Peace Corps: Wilfred Grenfell's Labrador Volunteers.* Bowie, MD: Heritage Books, 2001.

Tsing, Anna. *Friction: An Ethnography of Global Connection.* Princeton, NJ: Princeton University Press, 2004.

Tsing, Anna. *The Mushroom at the End of the World: On the Possibility of Life in Capitalist Ruins.* Princeton, NJ: Princeton University Press, 2015.

van Wyck, Peter C. *The Highway of the Atom.* Montreal: McGill-Queen's University Press, 2010.

Walker, Janet, and Nicole Starosielski, "Introduction: Sustainable Media." In *Sustainable Media*, edited by Nicole Starosielski and Janet Walker, 1–18. New York: Routledge, 2016.

"The Year at St. Anthony Hospital." *Among the Deep Sea Fishers* 25, no. 1 (April 1927): 19.

Young, Liam Cole. "Innis's Infrastructure: Dirt, Beavers, and Documents in Material Media Theory." *Cultural Politics* 13, no. 2 (2017): 227–49.

Zylinska, Joanna. *Nonhuman Photography.* Cambridge, MA: MIT Press, 2017.

www.ingramcontent.com/pod-product-compliance
Lightning Source LLC
Chambersburg PA
CBHW071738270326
41928CB00013B/2726